**NORTH CAROLINA
STATE BOARD OF COMMUNITY COLLEGES
LIBRARIES
ASHEVILLE-BUNCOMBE TECHNICAL COMMUNITY COLLEGE**

DISCARDED

JUN 19 2025

MATERIALS-HANDLING TECHNOLOGIES USED AT HAZARDOUS WASTE SITES

MATERIALS-HANDLING TECHNOLOGIES USED AT HAZARDOUS WASTE SITES

by

Majid Dosani
John Miller

PEI Associates, Inc.
Cincinnati, Ohio

NOYES DATA CORPORATION
Park Ridge, New Jersey, U.S.A.

Copyright © 1992 by Noyes Data Corporation
Library of Congress Catalog Card Number 91-46546
ISBN: 0-8155-1299-6
ISSN: 0090-516X
Printed in the United States

Published in the United States of America by
Noyes Data Corporation
Mill Road, Park Ridge, New Jersey 07656

10 9 8 7 6 5 4 3 2 1

Library of Congress Cataloging-in-Publication Data

Dosani, Majid.
    Materials-handling technologies used at hazardous waste sites / by Majid Dosani, John Miller.
        p.   cm. -- (Pollution technology review, ISSN 0090-516X ; no. 208)
    Includes bibliographical references (p.   ) and index.
    ISBN 0-8155-1299-6
    1. Hazardous waste sites--United States--Management. 2. Hazardous wastes--Government policy--United States. 3. Materials handling.
I. Miller, John C. II. Title. III. Series.
TD1040.D67   1992
628.4'2--dc20   91-46546
                                                  CIP

Foreword

This book summarizes the types of debris, material, and contaminants found at Superfund and other hazardous waste sites and the materials-handling equipment and general procedures used to perform site restoration and cleanup. It provides information on state-of-the-art materials-handling equipment and procedures useful for addressing difficult, site-specific, materials-handling problems.

The following factors affect the selection of equipment and procedures for materials handling at hazardous waste sites:

1. Type and quantity of contaminated materials present
2. Amount and type of contaminants found on-site
3. Type of removal or remedial action selected (capping, excavation, pumping, etc.)
4. Treatment processes implemented on-site
5. General site characteristics (climate, soil type and moisture, topography)

The book includes information concerning capabilities, performance, and applicability of a variety of materials-handling equipment and procedures at various hazardous waste sites and cost of their implementation. Case studies for 22 sites throughout the U.S. have been included to provide detailed information concerning debris, material, and contaminants found on-site; the specific materials-handling needs; and problems encountered. Additional information has been provided for 37 pieces of materials-handling equipment, including specifications, features, options, manufacturers, and photographs.

Because of the diversity of debris, material, and contaminants found on Superfund and other hazardous waste sites, each site must be evaluated individually for the selection and implementation of materials-handling equipment and procedures. Adequate published information concerning debris and material handled has not previously been available to environmental personnel and response contractors. Although attempts had been made to categorize debris and material present on-site, little information was found concerning their quantities at specific sites; and quantity estimates, which would be helpful to response contractors and environmental personnel for estimating cost of equipment operations, were lacking.

This book, then, is a compilation of information regarding materials-handling equipment and techniques that have been implemented at hazardous waste sites throughout the U.S. and Europe. The information will assist site remediation contractors and environmental personnel in evaluating materials-handling techniques for potential applicability at sites that require remediation.

The information in the book is from *Survey of Materials-Handling Technologies Used at Hazardous Waste Sites,* prepared by Majid Dosani and John Miller of PEI Associates, Inc. for the U.S. Environmental Protection Agency, June 1991.

The table of contents is organized in such a way as to serve as a subject index and provides easy access to the information contained in the book.

Advanced composition and production methods developed by Noyes Data Corporation are employed to bring this durably bound book to you in a minimum of time. Special techniques are used to close the gap between "manuscript" and "completed book." In order to keep the price of the book to a reasonable level, it has been partially reproduced by photo-offset directly from the original report and the cost saving passed on to the reader. Due to this method of publishing, certain portions of the book may be less legible than desired.

ACKNOWLEDGMENTS

This report was prepared for the U.S. Environmental Protection Agency, Office of Research and Development, Risk Reduction Engineering Laboratory, Cincinnati, Ohio, by PEI Associates, Inc. Majid Dosani, who served as PEI's Work Assignment Manager, and John Miller were the principal authors. Michael Taylor, Ph.D., was PEI's Project Manager and Jack Greber was Project Director. Other PEI staff contributing to the project were Richard Gerstle, Judy Hessling, Sherri Gill, Jerry Day, Martha Phillips, and Jim Scott.

Naomi P. Barkley served as the EPA Technical Project Monitor. The authors wish to thank Donald E. Sanning of EPA-RREL for providing technical assistance.

The authors also wish to thank the following industries and trade organizations for their help and cooperation:

- National Solid Waste Management Association (NSWMA), Washington, DC
- Montgomery County Landfill, VA
- National Aggregates Association, MD
- Dravo Corporation, OH
- Silver Hill Sand and Gravel, NJ
- Institute of Scrap Recycling Industries, Washington, DC
- National Association of Demolition Contractors, IL

NOTICE

The materials in this book were prepared as accounts of work sponsored by the U.S. Environmental Protection Agency. This information has been subject to the Agency's peer and administrative review and has been approved for publication. On this basis the Publisher assumes no responsibility nor liability for errors or any consequences arising from the use of the information contained herein. Mention of trade names or commercial products does not constitute endorsement or recommendation for use by the Agency or the Publisher.

The book is intended for information purposes only. The reader is cautioned to obtain expert advice before implementation of any procedures described, since materials to be handled are hazardous wastes. Final determination of the suitability of any information or procedure for use by any user, and the manner of that use, is the sole responsibility of the user.

All information pertaining to law and regulations is provided for background only. The reader must contact the appropriate legal sources and regulatory authorities for up-to-date regulatory requirements, and their interpretation and implementation. The book is sold with the understanding that the Publisher is not engaged in rendering legal, engineering, or other professional service. If advice or other expert assistance is required, the service of a competent professional should be sought.

Contents and Subject Index

1. **INTRODUCTION** 1
   Background 1
   Objectives 2
   Approach 2
   Report Organization 4

2. **CONCLUSIONS** 5

3. **RECOMMENDATIONS** 7

4. **SITE CHARACTERIZATION** 9
   Contaminant Profile 9
   Debris/Materials Found at Hazardous Waste Sites 9
   Equipment Used at Hazardous Waste Sites 11
   Equipment Availability 14

5. **MATERIALS-HANDLING EQUIPMENT AND PROCEDURES** 16
   Excavation and Removal 19
     Backhoe 19
       Capabilities 19
       Performance 21
       Application for Hazardous Waste Site Work 21
       Cost 24
     Front-End Loader 24
       Capabilities 24
       Performance 24
       Application for Hazardous Waste Site Work 24
     Crawler Tractors 28
       Capabilities 28
       Performance 29
       Cost 29
     Trencher 29

   Performance 29
   Application for Hazardous Waste Site Work 30
   Cost 30
  Skid-Steer Loader 30
   Performance 30
   Application for Hazardous Waste Site Work 30
   Cost 30
  Forklift Truck 30
   Performance 32
   Application for Hazardous Waste Site Work 32
   Cost 32
Dredging 32
 Mechanical Dredging 32
  Capabilities 32
  Performance 36
  Cost 36
 Hydraulic Dredging 38
  Capabilities 38
  Performance 38
 Pneumatic Dredges 38
  Capabilities 38
  Application for Hazardous Waste Site Work 40
  Cost 40
Pumping 40
 Application for Hazardous Waste Site Work 42
 Cost 43
Size and Volume Reduction 43
 Size Reduction 43
 Volume Reduction 44
Separation and Dewatering 46
 Component Separation 46
 Dewatering 47
  Thickening Agents 48
  Conditioning Agents 48
  Centrifuges 48
  Vacuum Filters 48
  Pressure Filters 48
  Dewatering Lagoons 48
Conveying Systems 49
Storage Containers, Bulking Tanks, and Containment 49
 Storage Containers and Bulking Tanks 49
 Containment Systems 51
  Vacuum Systems 51
Compaction 51
 Soil Stabilizer 51
 Equipment 52
Miscellaneous Equipment and Procedures 52
Drum Handling and Removal 53
Asbestos Remediation 55

Contents and Subject Index   ix

    Emission Control 56
    Low-Level Radioactive Waste 57
    Equipment Decontamination 58

6. **FOREIGN CONTACTS** 60

7. **CASE STUDIES** 64
    Western Sand and Gravel, Burrillville, Rhode Island (Region I) 65
    Iron Horse Park, Billerica, Massachusetts (Region I) 66
    Industrial Latex, Wallington, New Jersey (Region II) 67
    International Metallurgical Services, Newark, New Jersey (Region II) 69
    Bruin Lagoon No. 2, Bruin, Pennsylvania (Region III) 72
    Ambler Asbestos Tailings Pile, Ambler, Pennsylvania (Region III) 73
    Aberdeen Pesticide Site, Aberdeen, North Carolina (Region IV) 74
    A.L. Taylor Site, Brooks, Kentucky (Region IV) 75
    Midwest Plating and Chemical Corp., Logansport, Indiana (Region V) 77
    Aeroquip/Republic Hose, Youngstown, Ohio (Region V) 79
    G&H Landfill, Utica, Michigan (Region V) 81
    PBM Enterprises, Romulus, Michigan (Region V) 82
    Midco II, Gary, Indiana (Region V) 84
    Motco Site, La Marque, Texas (Region VI) 87
    Cleve Reber Site, Sorrento, Louisiana (Region VI) 88
    Quail Run, Gray Summit, Missouri (Region VII) 89
    Solid State Circuits, Republic, Missouri (Region VII) 92
    B&C Metals, Denver, Colorado (Region VIII) 94
    Burlington Northern Railroad, Somers, Montana (Region VIII) 95
    McColl Superfund Site, Fullerton, California (Region IX) 96
    Pacific Hide and Fur, Pocatello, Idaho (Region X) 98
    Northwest Transfer Salvage Yard, Everson, Washington (Region X) 99

REFERENCES 102

BIBLIOGRAPHY 105

**APPENDIX A: FREQUENCY OF OCCURRENCE OF CONTAMINANTS AT
1035 SUPERFUND SITES** 109

APPENDIX B: DEBRIS/MATERIALS CATEGORIZATION 116

**APPENDIX C: DEBRIS/MATERIALS CHARACTERIZATION FOR 100
HAZARDOUS WASTE SITES** 119

APPENDIX D: EQUIPMENT USED AT 100 HAZARDOUS WASTE SITES 125

**APPENDIX E: DEBRIS/MATERIAL-HANDLING OVERVIEW FOR 67
HAZARDOUS WASTE SITES** 152

APPENDIX F: EQUIPMENT COSTS FOR HAZARDOUS WASTE WORK 162

APPENDIX G: EQUIPMENT DESCRIPTIONS 165

GLOSSARY 204

COPYRIGHT NOTICE 210

ACRONYMS/ABBREVIATIONS 214

1. Introduction

BACKGROUND

Superfund and other hazardous waste sites in the United States contain many different types of materials that require physical separation, classification, and decontamination. These various materials are often contaminated with hazardous chemical residues. In some instances, however, although a material found on-site contains no hazardous substance, it still must be handled and/or disposed of offsite (e.g., transported to a sanitary landfill).

A typical Superfund or other hazardous waste site contains hazardous chemicals that are frequently mixed with the remnants of razed structures (wood, steel, concrete blocks); municipal and/or industrial solid wastes; metallic debris (refrigerators, abandoned cars, drums, transformer casings); and contaminated soils, sludges, and liquids. Materials-handling and classifying technologies are needed to deal with the large quantities of these various materials prior to, or in conjunction with, their decontamination and disposal.

For the purposes of this study, debris has been defined as any unused, unwanted, or discarded solid or liquid that requires staging, loading, transporting, pretreating, treatment, and/or disposal on a hazardous waste site. In addition to debris, other materials (e.g., soil, sludge, asbestos, and various liquids) must also be handled.

Materials-handling procedures may also be required for other types of activities that occur at a hazardous waste site. Site preparation may require the use of heavy equipment to build access roads, containment trenches, or concrete decontamination pads. Pretreatment processes requiring materials-handling techniques (screening, size reduction, dewatering) may be necessary before a remedial treatment technology is applied. Many current sites do not require the cleanup of hazardous waste; instead, they are essentially construction projects involving the building of treatment facilities or new wells for dealing with groundwater contamination of unknown origin. Such projects may require a variety of materials-handling equipment and procedures not normally encountered during cleanup operations (e.g., the laying of pipe or deep-well excavation techniques).

To date, published information concerning materials handling at hazardous waste sites is sparse. An engineering and economic analysis is needed to develop a data base useful to EPA personnel and response contractors requiring equipment or procedures to address difficult materials-handling problems.

This document provides technical information for the selection and implementation of materials-handling equipment and procedures as they relate to remediation of hazardous waste sites.

OBJECTIVES

The objectives of this study are as follows:

- Characterize Superfund and other hazardous waste sites in terms of frequently found contaminants and on-site materials that need to be handled or decontaminated.

- Compile a summary of equipment and processes used for the handling, separation, and decontamination of debris and material at domestic and foreign hazardous waste sites.

- Obtain descriptions from industry and vendors of materials-handling equipment that may be adaptable for use at hazardous waste sites.

- Present detailed case studies illustrating both typical and atypical materials-handling problems encountered at hazardous waste sites.

- Provide information on state-of-the-art materials-handling equipment and procedures that EPA personnel and response contractors might find useful for addressing difficult, site-specific, materials-handling problems.

APPROACH

In this study the following approach was used to summarize the types of debris/material and contaminants found at Superfund and other hazardous waste sites as well as the materials-handling equipment and general procedures that have been implemented to perform site remediation:

1) A comprehensive literature search of 75 sources was conducted. These sources covered the following subjects:

    - General materials-handling equipment and procedures
    - Process equipment for resource recovery
    - Equipment/construction cost guides
    - Onsite materials handling
    - Contaminants profiles
    - Miscellaneous
        - Stabilization/solidification
        - Dust control
        - Physical, chemical, and biological treatment
        - Case studies
        - Determining compatibility of hazardous waste
        - SITE program

2) Forty-five Records of Decisions (RODs) were reviewed from sites in Regions I, III, and V to provide background site descriptions for contaminants and material found on site.

3) On-Scene Coordinator (OSC) reports for 94 sites in 9 of the 10 EPA Regions were reviewed for specific materials-handling equipment, procedures, and/or problems encountered during site remediation. OSC Reports were not available from Region IX immediately after the 1989 San Franscisco earthquake.

4) Representatives from various industries and their respective trade unions were contacted to obtain information on equipment/procedures that might be adapted for hazardous waste site work. These contacts included:

   o Landfills
   o Junk dealers (scrap metal)
   o Sand and gravel companies
   o Demolition contractors
   o Mining industry

5) Nine computerized data bases were investigated for site characterization and equipment used:

   o Record of Decision Data Base (ROD)
   o National Priorities List Data Base (NPL-Mitre Corp.)
   o Hazardous Waste Data Management System (HWDMS)
   o Resource Conservation and Recovery Information Service (RCRIS)
   o Comprehensive Environmental Response, Compensation, and Liability Information System (CERCLIS)
   o Superfund Site Tracking Information System (SUPTRK)
   o Hazardous Waste Data Base (HAZARD)
   o Technical Information Exchange (COLIS)
   o National Technical Information Service (NTIS)

6) Emergency Response Cleanup Services (ERCS) files for Region V (Zone 3) were reviewed for site descriptions, including equipment used for site remediation.

7) The OSCs from Regions I, II, III, and V, ERCS personnel, and response contractors from Region V were all contacted for site-specific information regarding materials-handling efforts.

8) Vendors were contacted for information concerning the capability and cost of materials-handling equipment.

9) Equipment and processes that have been implemented for handling, separating, and decontaminating materials at hazardous waste sites outside the United States were obtained by contacting North American Treaty Organization/Committe on Challenges of Modern Society (NATO/CCMS) Pilot Study participants as well as representatives from companies conducting onsite remedial activities in Germany, the Netherlands, France, and the United Kingdom.

REPORT ORGANIZATION

This report has been designed to provide quick, handbook-type access to various equipment and procedures used for hazardous waste site remediation. Section 4 addresses site characteristics, including contaminants present, the types of debris and material found on selected sites, and equipment used. This section also presents information concerning response contractor equipment inventories. Section 5 provides in-depth reviews of materials-handling equipment and procedures. This section also includes an evaluation of the capabilities, performance, and application of this equipment for handling materials at hazardous waste sites. Cost information for renting, leasing, or purchasing such equipment is also provided. Section 6 presents case studies of 22 hazardous waste sites. These case studies include site descriptions, contaminants present, equipment used for site remediation, specific materials-handling problems, and the ultimate disposition of contaminated material. Appendix E contains an overview of the debris/material handled, major contaminants and principal materials-handling equipment or procedure used for 67 hazardous waste sites. Appendix G contains examples of equipment that have application for hazardous waste site remediation.

2. Conclusions

The following conclusions were reached during this study:

1) The diversity of debris, material, and contaminants found on Superfund and other hazardous waste sites requires that evaluations be made on an individual, site-specific basis for the selection and implementation of materials-handling equipment. Factors affecting the selection of equipment and procedures for materials handling are as follows:

    ° Type and quantity of contaminated materials present
    ° Amount and type of contaminants found on-site
    ° Type of removal/remedial action selected (capping, excavation, pumping, etc.)
    ° Treatment process(es) implemented on-site
    ° General site characteristics (climate, soil type and moisture, topography)

2) No technical data base currently exists that covers both the characterization of the debris and material to be handled and the equipment and procedures used to deal with these materials at hazardous waste sites. References to debris/material found and equipment used are random and often fall under the heading of "miscellaneous".

3) Equipment and procedures used for materials handling (e.g., sand/gravel, demolition, etc.) have not yet been adequately explored to take advantage of innovations that could benefit hazardous waste site remediation. The hazardous waste field has need for new devices as well as utilizing existing equipment not now being used to handle hazardous waste and related materials.

4) Existing available information concerning debris and materials handled is inadequate for EPA personnel and response contractors. Attempts have been made to categorize the debris and materials that have been found on-site, but virtually no information is available concerning the quantities of debris and materials handled. Estimates of quantities would be helpful to response contractors and EPA personnel for disposal costs estimations.

5) Based on the information gathered from contacts in Germany, the Netherlands, France, and the United Kingdom, materials-handling equipment and procedures currently used for remediation of foreign sites are essentially the same as those used in the United States. Extensive hazardous waste site work is currently being conducted in these countries, however, and contact should be maintained to monitor future development of any new or innovative materials-handling technologies.

6) Data concerning materials-handling equipment and procedures used to remove or remediate contaminated debris and materials at hazardous waste sites are difficult to access for most of the Regions. Files containing information regarding the equipment used on-site are often unavailable. Also, much of the equipment information is tied in with cost information and therefore subject to confidential business information (CBI) restrictions, which require a Freedom of Information form requesting access, to be filled out prior to obtaining the information.

7) Based on the information gathered for this report, the type of contaminant (e.g., acids, low-flash point liquids) does not appear to have a direct effect on the choice of equipment and procedures used at hazardous waste sites. Whereas the type of contaminant found on site affects the level of personal protection required, the selection of most equipment is based on cost, availability, and the ability to deal with the physical nature of the debris/material to be handled. The contaminant type on-site does, however, affect modifications to the equipment chosen for the site work (e.g., splash shield installation on excavation equipment).

3. Recommendations

The results of this study have prompted the following recommendations for further action:

1) Given the wide diversity of debris and materials found on hazardous waste sites and the different methods of site remediation used, a central computerized data base dealing solely with on-site materials handling should be compiled. This would provide EPA personnel and response contractors with a starting point for dealing with specific materials-handling problems. A computerized data base could also be updated on a regular basis to provide for changes in the field of materials-handling, or to provide site-specific information for on-going remediation/removal projects involving similar conditions (e.g., landfills, battery breaking operations, etc.) Workshops for all 10 EPA Regions should be developed to brief interested personnel about the content, availability, and accessing of the data base.

2) Vendors of equipment with hazardous waste site applications should be alerted to potential opportunities for hazardous waste site work. Workshops/seminars sponsored by the U.S. EPA could be offered to vendors. Additionally, U.S. EPA could develop and implement an international conference on materials-handling.

3) A standard operating procedure (SOP) should be developed for dealing with materials found on hazardous waste sites that have the potential to be recycled or reclaimed. A more comprehensive analysis should be performed to investigate the feasibility of using recyclable materials found on-site (after classification and decontamination). For example, many sites contain large amounts of scrap metal that could be salvaged after being decontaminated.

4) Recommended followup studies are as follows:

    ° Development of a computerized materials-handling data base and electronic bulletin board containing information for both domestic and foreign hazardous waste sites.

    ° Development and implementation of seminars/training sessions dealing with difficult, site-specific, materials-handling problems. These seminars/training sessions would be conducted to keep RPMs and OSCs updated on new and innovative pieces of equipment and available options/accessories. Field demonstrations of equipment and procedures for dealing with debris/materials found on hazardous waste sites would also be conducted.

- Continued monitoring of future development of new or innovative materials-handling technologies being used at domestic or foreign hazardous waste sites (e.g., U.S. EPA recently developed and demonstrated a pilot-scale debris washer for handling contaminated metal and concrete found on hazardous wastes sites).

- Collection of additional information concerning the effect of equipment downtime and parts availability on overall project costs and schedules.

4. Site Characterization

CONTAMINANT PROFILE

Superfund and other hazardous waste sites contain a variety of hazardous chemicals. Concentrations of these contaminants can range from several parts per billion of a single compound to thousands of drums containing high concentrations of complex mixtures of organic chemicals. Several studies have attempted to characterize the contaminants found on National Priorities List (NPL) sites. The U.S. EPA has developed first- and second-priority lists of hazardous substances most commonly found at NPL facilities that have been determined to pose a significant potential threat to human health, as required under SARA [52 Federal Register (FR) 12866, April 17, 1987, and 53 FR 41280, October 20, 1988]. Significant compounds are also listed under the Resource Conservation and Recovery Act (RCRA) (Appendix IX), Comprehensive Environmental Response Compensation and Liability Act (CERCLA) (Superfund Contract Laboratory Program), and Clean Water Act (CWA) (Priority Pollutant Compounds). The U.S. EPA and State environmental agencies may add or delete substances from these lists as well as characteristics used to identify hazardous wastes not on the lists. Table 1 presents a list of the 25 most commonly encountered contaminants found at Superfund sites (Pasha Publication Co. 1989). Appendix A contains a complete list of hazardous substance found at 1035 Superfund sites. Also contained in the appendix is a breakdown of the frequency of occurrence of contaminants by U.S. EPA region.

The wide variety of chemicals found on hazardous waste sites may lead to handling problems because of factors such as high corrosivity (resulting in a need for corrosion-resistant equipment) or highly toxic volatile compounds, which require special personal protective equipment. Additional handling problems also may arise from reactions occurring because of the complex mixtures of chemicals found on site.

DEBRIS/MATERIALS FOUND AT HAZARDOUS WASTE SITES

Tamm, Cowles, and Beers (1988) defined debris categories based on information derived from interviews with EPA personnel and response contractors. The nine categories of debris presented in that report, however, were in the context of feedstock preparation. As defined for this report debris is any unused, unwanted, or discarded solid or liquid that requires staging, loading, transporting, pretreating, treatment, and/or disposal on a hazardous waste site. These three researchers estimated that debris occurring on hazardous waste sites ranges in quantity from less than 1 percent to greater than 80

TABLE 1. FREQUENCY OF OCCURRENCE OF CONTAMINANTS FOUND AT 1035 SUPERFUND SITES (1989)[a]

1.	Trichloroethylene	246
2.	Lead	230
3.	Chromium	173
4.	Polychlorinated Biphenyls (PCBs)	156
5.	Heavy Metals	147
6.	Tetrachloroethylene	138
7.	Benzene	137
8.	Toluene	131
9.	Volatile Organic Compounds (VOCs)	129
10.	Arsenic	119
11.	Cadmium	100
12.	1,1,1-Trichloroethane	86
13.	Copper	74
14.	Zinc	71
15.	Vinyl chloride	68
16.	Xylene	67
17.	Chloroform	65
18.	Phenols	64
19.	1,1-Dichloroethane	60
20.	Waste Solvents	57
21.	Cyanides	53
22.	Nickel	46
23.	1,1-Dichloroethylene	45
24.	Ethyl Benzene	45
25.	Methylene Chloride	45

[a] Source: 1989 Guide to Superfund Sites.

percent of the total waste found on site. In addition to debris, other
materials (e.g., soil, sludge, asbestos, and various liquids) must be handled.
Debris and other material that require special materials-handling, are
presented in the following 12 general categories:

- Textiles
- Glass
- Paper
- Metal
- Plastic
- Rubber
- Wood/vegetation
- Construction debris
- Soil
- Sludge
- Liquids
- Asbestos

Appendix B presents a detailed breakdown of these 12 categories.

A review of information from 100 hazardous waste sites in all 10 EPA
Regions to identify the frequency of occurrence of debris or materials that
required materials-handling was vague. One of the problems encountered in
categorizing the debris is that most RODs and OSC reports refer to "miscellaneous debris" and give no significant details about debris/material handling. Figure 1 presents the frequency of occurrence for each of the debris/material types found on 100 waste sites. It should be noted that this list
represents what was mentioned--not what might actually have been on site.
Appendix C offers a more detailed breakdown of the 100 sites surveyed for
debris/material. It should be noted that the debris/material profiles at
hazardous waste sites are relatively constant nationwide. Contaminated soil,
liquids, and metals (drums, etc.) are the materials most commonly encountered
that require handling at the sites investigated for all 10 Regions. No
significant difference in debris and material appears to exist among the
sites found in the various Regions.

EQUIPMENT USED AT HAZARDOUS WASTE SITES

Information from 100 hazardous waste sites was reviewed to obtain a
profile of equipment used for materials-handling on site. Table 2 shows the
number of sites at which each particular piece of equipment was used. Appendix D contains a breakdown of the exact equipment needs at the 100 sites.
Projects at the reviewed sites ranged from major removal/excavation efforts
involving sludge, soil, drums, tanks, and liquids to simple removal actions
of only a few drums.

Cross referencing of Appendices C and D gives an indication of the
equipment needed to deal with various debris/material categories. In addition
to the equipment and debris/material profiles presented in Appendices C and
D, Appendix E contains an overview of 67 sites from all 10 Regions, which
indicates contaminant, debris/material handling, and the primary debris-handling procedures and equipment implemented.

The major point that emerged from discussions with EPA personnel and
response contractors involved on-site remediation is that equipment usage/modification/fabrication is site-specific and often involves trial and error.

12 Materials-Handling Technologies Used at Hazardous Waste Sites

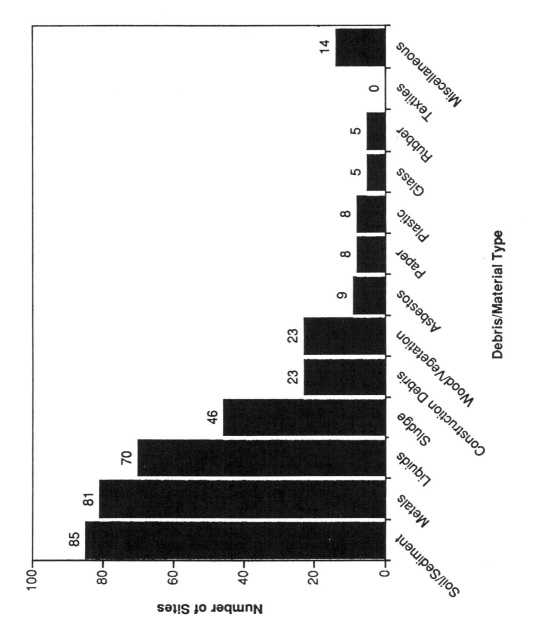

Figure 1. Frequency of occurrence for types of debris/materials found on 100 hazardous waste sites.

TABLE 2. EQUIPMENT USED FOR MATERIALS HANDLING AT 100 HAZARDOUS WASTE SITES

Equipment	Number of sites
Excavation/removal	
Backhoe/excavator	76
Front-end loader	62
Lowboy	46
Bulldozer	42
Tractor (OTR)	32
Skid steer loader	24
Forklift	21
Crane	21
Grader	9
Dragline	1
Pump/vacuum unit	
Diaphragm pump	36
Vacuum truck	21
Submersible pump	16
Trash pump	15
Vacuum unit	15
Barrel pump	7
Separation/size reduction	
Crusher (drum/debris)	10
Shredder (tire, drum)	9
Vibrating screen	3
Conveyor	2
Miscellaneous	
Generator	42
Hand tools (shovels, hammers, etc.)	41
Pressure washer/laser	37
Air compressor	33
Bulking tanks/pools	26
Rolloff boxes	25
Drum grappler	21
Cutting torch	20
Drum cart	16
Nonsparking tool set	14
Chain saw	13
Air hammer	13
Drum punch	8
Hoe ram/pile driver	4
Pug mill	1

Despite the site-specific nature of hazardous waste remediation, similar conditions are often found at many sites (i.e., landfills, battery breaking operations etc.). This results in similar techniques and SOPs being used at different sites for materials-handling. The following are examples of site-specific solutions to problems involving equipment:

- Hydraulic systems had to be modified to adapt a backhoe for drum handling (grappler).

- Rubber or foam tires instead of pneumatic tires were used at sites with large quantities of sharp metal/glass objects.

- Splash shields had to be installed on heavy equipment.

- Larger bulldozers were used to winch smaller dozers up and down the steep grades of asbestos tailings piles.

- Propane-powered instead of diesel-powered loaders were used for inside work to reduce fumes.

- Heavy equipment failed due to weather (e.g., cracked hydraulic lines from cold, tractability during icy conditions, metal fatigue from digging in frozen soil).

- A drum crusher instead of backhoe was used to crush drums.

- Rolloff boxes were converted into treatment chambers for cyanide-contaminated film chips.

Detailed descriptions of how materials-handling equipment and procedures were implemented at each of 22 sites are contained in Section 7, Case Studies.

EQUIPMENT AVAILABILITY

Response contractor equipment inventories vary according to the size of the company and type of site remediation typically performed. In most cases contractors maintain a standard inventory and rent or lease large, more expensive pieces of equipment (e.g., bulldozers, cranes) for site-specific needs. Some of the larger, more specialized contractors maintain inventories that include heavy equipment. Table 3 presents a list of standard inventory equipment and frequently rented or leased equipment.

TABLE 3. LIST OF STANDARD INVENTORY EQUIPMENT VS. RENTED/LEASED EQUIPMENT FOR RESPONSE CONTRACTORS

Equipment	Inventory	Rented/leased
Backhoe		X
Front-end loader		X
Bulldozer		X
Lowboy		X
Forklift		X
Crane		X
Gradall		X
Skid steer loader		X
Diaphragm pump		X
Trash pump	X	
Submersible pump	X	
Vacuum unit	X	
Vacuum truck		X
Barrel pump		X
Drum crusher		X
Vibrating screen		X
Tire shredder		X
Hand tools	X	
Nonsparking tools	X	
Pressure washer		X
Generator	X	
Drum grappler		X
Air compressor		X
Bulking tanks/pools	X	
Air hammer		X
Chain saw	X	
Cutting torch		X
Barrel cart	X	
Drum punch		X

5. Materials-Handling Equipment and Procedures

Materials handling can be defined as "...a system or combination of methods, facilities, labor, and equipment for moving, packaging, and storing of materials to meet specific objectives" (Kulwiec 1985). Within the context of hazardous waste site remediation, materials handling can encompass everything from site preparation (e.g., the building of access roads) to the actual treatment processes. In general, onsite materials-handling equipment and procedures are used for the following purposes:

- Physical separation and classification
- Site preparation
- Construction
- Feedstock preparation and handling
- Equipment, structure, and building decontamination
- Loading and hauling

The selection of equipment for materials handling will usually be a function of several important site-specific considerations. These include (Cullinane, Jones, and Malone 1986):

- General site characteristics (vegetation, soil type and moisture, topography)
- Quantity of material present
- Treatment technology implemented (pretreatment needs)
- Debris characteristics (metal, plastics, construction, etc.)
- Waste characteristics (solid, liquid, sludge)
- Packaging of waste materials (drums, tanks, lagoons, etc.)
- Ease of startup and demobilization
- Climate (temperature, precipitation)
- Size of working area

Figure 2 presents a flow diagram of general materials-handling procedures encountered on hazardous waste sites. The suitability of a particular piece of equipment for general onsite use will depend on the following factors (Doerr, Landin, and Matrin 1986):

- Cost
- Availability
- Personnel requirements for operation, maintenance, and safety
- Versatility
- Storage requirements
- Objectives of treatment

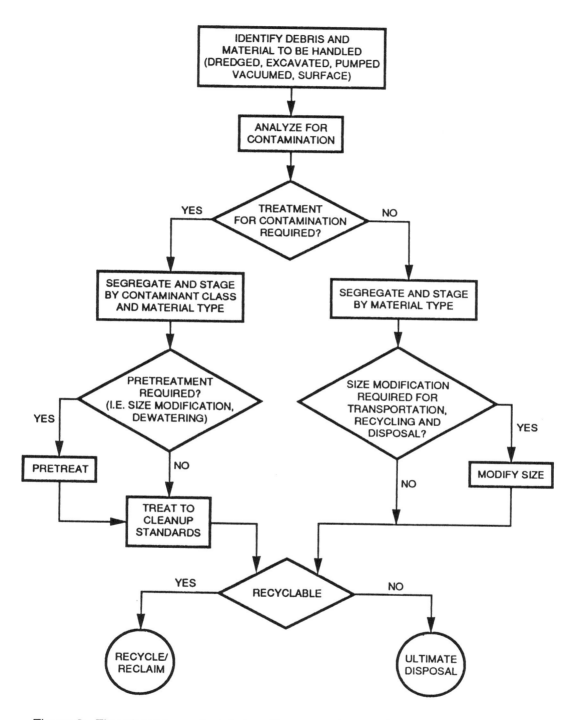

Figure 2. Flowsheet for materials-handling procedures at hazardous waste sites.

When equipment requirements have been determined, the job cost for each piece of equipment can be estimated. Equipment can be purchased, rented, or leased. Rented or leased equipment is initially less expensive to use, but the cost of purchased equipment can be amortized over several projects.

Cost estimates for several pieces of equipment are provided herein, and an overall summary of costs is presented in Appendix F. Detailed cost estimates for operating, renting, leasing, and purchasing equipment can be obtained from Church's (1981) "Excavation Handbook" and R.S. Means (1988) "Building Construction Cost Data." The cost estimates were obtained from response contractors and equipment vendors. Routine maintenance (e.g., oil, diesel fuel) costs have been figured into the rent/lease costs presented in Appendix F. The rental/lease information listed in Appendix F was obtained from response contractors in EPA Region V; therefore, actual costs may vary for other EPA Regions.

Frequently used materials-handling procedures at hazardous waste sites can be categorized as follows:

- Excavation and removal
- Dredging
- Pumping
- Size and volume reduction
- Separation and dewatering
- Conveying systems
- Storage containers, bulking tanks, and containment
- Drum handling and removal
- Compaction
- Miscellaneous equipment and procedures
- Asbestos remediation
- Handling of low-level radioactive waste
- Emission control
- Equipment decontamination

A wide range of equipment is available for conducting each of these procedures. One or more pieces can be selected as needed.

The remainder of this section presents detailed technical information (including capability, performance, and hazardous waste site applications) for various materials-handling equipment and procedures. Appendix G also contains examples of specific models of equipment that have been used or have potential for use on hazardous waste sites. Included in this appendix are:

- Photographs
- Specifications
- Features
- Attachments
- Options
- Manufacturers

Finally, representatives from several industries (e.g., sand and gravel, demolition scrap dealers) that regularly use materials-handling equipment were contacted to obtain specific information concerning equipment usage and modifications that might be applicable for hazardous waste site work. Although these industries use a variety of materials-handling equipment, most of the companies contacted were unwilling to provide any information.

EXCAVATION AND REMOVAL

Most of the equipment used for excavation and removal work at hazardous waste sites is standard heavy construction equipment. Selection of excavation equipment depends on the quantity and physical properties of the debris and materials present. Table 4 presents excavation equipment performance characteristics. Because the materials found on sites vary, selection of excavation equipment is very site-specific. Excavation or removal processes take place at most sites. As shown in Table 2, backhoes are the most commonly used equipment. Excavation techniques are most applicable for dealing with solid and thickened sludge materials. Conventional excavation techniques are less suitable when debris/materials have a high liquid content.

In addition to the characterization of the debris and materials to be handled, several other important factors must be considered before the excavation or removal of materials from a site. The selection of equipment (type and size) may depend on one or all of these factors, including (Wagner et al. 1986):

- Density of the waste on a site (average densities of landfilled wastes are reported to be from 800 to 1000 lb/yd^3 with moderate compaction) (Brunner and Keller 1972).
- Settlement of the fill.
- Bearing capacity of the site.
- Decomposition rate of wastes present.
- Packaging of the waste (drums, tanks, etc.).

Excavation equipment generally operates in a batch rather than continuous mode. This aspect has the advantage of being able to deal with localized areas of contamination within a hazardous waste site. Excavation and removal equipment is applicable under virtually all site conditions; however, such application may be cost-prohibitive at great depths or under varied hydrogeologic conditions. The capacities, horsepower, and size of equipment used for excavation and removal vary widely. Also, a variety of attachments, accessories, and options are available for individual pieces of equipment (as shown in Appendix G). The following subsections deal with various pieces of excavation and removal equipment that have proved to be applicable for work at hazardous waste sites.

Backhoe

Capabilities--
The backhoe, an excavator for subsurface work, is useful for trench digging and area excavation. It may be crawler-mounted (trackhoe) or wheel/tire-mounted. Machines having a capacity of greater than 2 yd^3 are generally

TABLE 4. EXCAVATION/REMOVAL EQUIPMENT PERFORMANCE CHARACTERISTICS[a]

Excavation/ removal equipment	General excavation	Ability to excavate hard and compacted material	Ability to excavate low-solids material	Soil hauling	Sludge hauling	Mixing of solids, soils	Spreading cover	Site maneuverability	Debris hauling
Wheel-mounted backhoe	A	A	B	B	B	A	A	A	A
Crawler-mounted backhoe	A	A	B	0	B	A	A	B	B
Wheel-mounted front-end loader	A	A	B	A	A	A	A	A	A
Crawler-mounted front-end loader	A	A	B	B	B	A	A	B	B
Skid steer loader	A	B	B	B	B	A	B	A	A
Bulldozer	A	A	0	0	0	0	A	B	0
Forklift truck	0	0	0	0	0	0	0	A	A
Dragline	A	A	A	0	0	0	0	0	0

[a] A = Good choice. Equipment is fully capable of performing function listed.
B = Secondary choice. Equipment is marginally capable of performing function listed.
0 = Not applicable or poor choice.

crawler-mounted to aid in stability. Smaller backhoes, which may be crawler- or wheel-mounted, are used for exploratory or smaller excavation jobs (Church 1981). A backhoe frequently used for hazardous waste work is a wheel-mounted combination backhoe and front-end loader (Figure 3). Several different pieces of auxiliary equipment are available for backhoes, including clamshell buckets, drum grapplers, dippers, loader buckets, and air percussion hammers. The term "backhoe" was used in many of the equipment logs examined when referring to both type backhoes (containing both a hoe and front-end load bucket) and excavators (having a hoe only). Excavators are usually larger and have a greater net horsepower, digging depth, reach, and bucket size.

Table 5 shows the maximum reach and depth of various-sized backhoe buckets. Table 6 shows the potential hourly production rates of a hydraulic backhoe under a variety of conditions. Table 7 gives specifications for crawler-mounted and diesel-engine-driven backhoes.

Performance--

Backhoes, the most common materials-handling equipment used for excavation, have wide application for most categories of debris/material. They are used extensively for solids and sludge excavation removal. When used with various pumping or vacuum units, they are applicable for low- solids straining. The backhoe bucket is easily controlled for precise width and depth excavation. Backhoes can excavate both hard and compacted materials. Tracked backhoes can traverse a variety of terrains with little difficulty; wheeled backhoes need a more level terrain for stability.

Disadvantages of the backhoe include limited digging depth (45 feet maximum), a linear reach of only 100 feet, and limited capabilities for backfilling or compacting (JRB Associates 1982). Onsite decontamination of backhoes, as with most heavy equipment, is generally accomplished with high-pressure hot water washers.

Application for Hazardous Waste Site Work--

Backhoes are readily available and easily transportable. They can be rented or leased in most areas. They are also the most versatile piece of heavy equipment for this application. Although used primarily for excavation and trenching, other on-site applications include the following:

- Drum sampling with a drum plunger attachment
- Drum crushing (generally used on sites with less than 150 drums to be crushed)
- Offloading of equipment from trucks (generally wheel-mounted backhoes with bucket attachment)
- Staging and/or loading of drums or debris with bucket attachment (structurally sound drums)
- Mixing of contaminated soil with various stabilization/solidification agents
- Drum handling with a hydraulic drum grappler attachment (requires modification of backhoe hydraulic system)

Sources: PEI Associates, Inc. 1990

Figure 3. Typical wheel-mounted backhoe.

TABLE 5. MAXIMUM REACH AND DEPTH OF VARIOUS-SIZED BACKHOES[a]
(maximum digging angle of 45 degrees)

Hoe size, yd^3	Maximum reach of boom, ft	Maximum depth of excavation, ft
1	35	22
1.5	42	25
2	49	30
3.5	70	45

[a] Source: Ware and Jackson 1978.

TABLE 6. THEORETICAL HOURLY PRODUCTION OF A HYDRAULIC BACKHOE[a]
(yd^3/h)

Product	Bucket size, yd^3						
	1	1.5	2	2.5	3	3.5	4
Moist loam, sandy clay	85	125	175	220	275	330	380
Sand and gravel	80	120	160	205	260	310	365
Common earth	70	105	150	190	240	280	330
Hard, dense clay	65	100	130	170	210	255	300

[a] Source: Ware and Jackson 1978.

TABLE 7. REPRESENTATIVE SPECIFICATIONS FOR
CRAWLER-MOUNTED AND DIESEL-ENGINE-DRIVEN BACKHOES[a]

Specification	Nominal dipper capacity, cysm[b]					
	1	2	3	4	5	6
Working weight, 10^3 lb	59	113	174	201	264	352
Horsepower for backhoe work	110	146	205	250	300	360
Range of dipper sizes for rock excavation, yd^3/m^2	7/8 to 1	1-3/4 to 2	2-1/2 to 3	3-1/4 to 4	4-1/4 to 5	5-1/4 to 6
Digging radius, ft	38	47	51	56	60	64
Digging depth, ft	23	29	32	36	40	44
Dumping height, ft	12	17	19	21	22	24

[a] Source: Church 1981 (reproduced with permission).
[b] Cubic yard struck measurement (see Glossary).

Safety modifications mentioned for hazardous waste work include installation of splash shields and installation of compressed-air units on the back of the equipment to allow the operator to work for longer periods of time on sites that require supplied air. One company has recently marketed a tele-operated remote-controlled excavation system (wheel-mounted) (Deere and Co. 1989). This unit can be used for excavating, bulldozing, heavy lifting, concrete breaking, or the remote handling of low-level radioactive waste. It can be operated up to a mile away via radio transmission, coaxial cable, or fiber-optic cable which provides maximum worker safety. A cab-mounted camera allows for overall or close-up viewing of the work area. Figure 4 shows a remote-operated excavator in use.

Cost--
Appendix F presents rent/lease/purchase cost summary information for backhoes. Purchase prices for wheel-mounted backhoes range from $28,000 to $90,000, depending on size and the attachments included. Purchase prices for crawler-mounted backhoes range from $100,000 to $650,000, depending on size.

Front-End Loader

Capabilities--
Front-end loaders may be crawler- or wheel-mounted; bucket capacities range from 1 to 5 yd^3 and 1 to 20 yd^3, respectively (Church 1981). Figure 5 shows operating dimensions and bucket action for a typical medium-sized bucket loader. Front-end loaders are equipped with buckets for digging, lifting, dumping, and hauling materials (U.S. EPA 1985a).

Performance--
Front-end loaders have a wide application for most categories of debris/materials. Because of their excellent flotation and traction capabilities, crawler-mounted loaders are ideal for unstable uneven terrain. Depending on the type of tires used, wheel-mounted loaders can maneuver on rough, muddy, and sloping terrain. Wheel-mounted loaders are faster and more mobile than crawler-mounted machines (U.S. EPA 1985a).

Front-end loaders are suitable for the excavation, spreading, and compaction of cover materials. They are excellent for hauling over short distances (less than 300 ft). The disadvantages of front-end loaders include their unsuitability for hauling long distances and they often must be used in conjunction with other types of excavation equipment (Noble 1976). Decontamination is generally accomplished by use of high-pressure hot-water washers.

Application for Hazardous Waste Site Work--
Front-end loaders are readily available and are easily transportable. Onsite application of front-end loaders includes the following:

- Hauling and staging of drums using bucket
- Offloading of equipment from trucks
- Staging of debris/materials (Figure 6)
- Feeding materials into treatment processes (e.g., shredding)
- Drum crushing (generally less than 150 drums) using bucket

Materials-Handling Equipment and Procedures 25

Figure 4. Remote operated excavator.

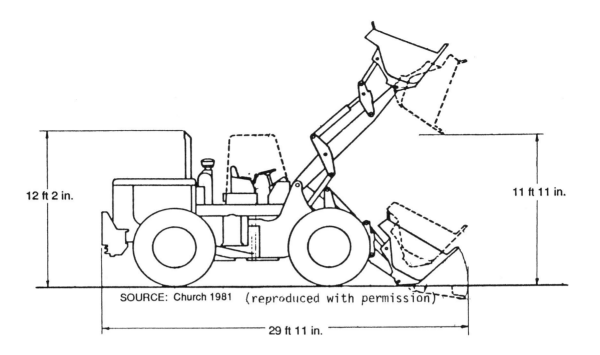

Figure 5. Operating dimensions and bucket action of a bucket loader.

Materials-Handling Equipment and Procedures 27

Source: PEI Associates, Inc. 1990

Figure 6. Front-end loader being used for debris removal at a hazardous waste site.

- Soil and sludge excavation
- Site preparation (access road construction)

Safety modifications mentioned were similar to those for backhoes (i.e., safety shields and compressed-air units).

Cost--

Appendix F presents cost summary information for front-end loaders. Renting/leasing rates vary according to the size of the equipment, the supplier, and the geologic area. The purchase prices of crawler-mounted loaders range from $24,000 for smaller (0.75-yd^3) buckets to $250,000 for larger (4.5-yd^3) buckets. Purchase prices of wheel-mounted units range from $13,000 (0.75-yd^3) to $130,000 (3.5-yd^3).

Crawler Tractors

Capabilities--

Crawler tractors (bulldozers) are generally equipped with a hydraulically controlled (vs. a mechanical cable hoist) blade and bucket lift, and they are usually crawler-mounted (U.S. EPA 1985a). They have a variety of blades (straight U-shaped, angle-type), which are used for the following purposes (Doerr, Landin, and Matrin 1986):

- Multipurpose blades with digging teeth to cut trenches and to uproot trees, shrubs, and rocks
- Pushing blades to uproot and fell large trees
- Cutting blades to shear off trees and shrubs at or below soil surface
- Stacking rakes, which are adapted for clearing

Table 8 presents specifications for crawler-tractor bulldozers with straight U-shaped blades.

TABLE 8. REPRESENTATIVE SPECIFICATIONS FOR CRAWLER-TRACTOR BULLDOZERS WITH STRAIGHT U-SHAPED BLADES[a]

Tractor engine horsepower	Blade data			
	Weight, lb	Height, ft	Width, ft	Capacity, cylm[b]
105	24,600	3.2	8.7	2.0
195	49,800	4.8	11.8	4.8
300	75,000	5.0	13.9	7.6
410	97,200	6.0	14.4	11.4
524	146,000	7.1	17.0	18.9
700	173,100	7.0	19.8	21.3

[a] Source: Church 1981 (reproduced with permission).
[b] Cubic yard loose measurement (see Glossary).

Performance--
Bulldozers are applicable for a variety of waste types. Bulldozers with variable cleat design are generally very stable; however, they are susceptible to sliding during icy weather. They are frequently used for excavating cover materials, clearing debris, constructing temporary or permanent dikes and containments, and compaction. Disadvantages of bulldozers include limited speed, limited mobility, and poor hauling capabilities. Decontamination is generally accomplished by use of high-pressure hot-water washers.

Application for Hazardous Waste Site Work--
Bulldozers are readily available and easily transportable. Although availability is not a concern, the size of the working area at hazardous waste sites may limit the usefulness of bulldozers. Applications for the use of bulldozers include:

- Drum excavation
- Site preparation, excavation, clearing of access roads, and clearing of vegetation
- Mixing of contaminated soil with stabilization/solidification agents
- General earth moving for capping of landfills

Safety modifications include installation of splash shields and the addition of compressed-air units for extended working periods.

Cost--
Cost summary information for bulldozers is presented in Appendix F. The most common sizes used for site work are the CAT D-3 and D-6 (or equivalent) units. The purchase price of a 65-hp (CAT D-3) bulldozer is approximately $50,000. The larger, 140-hp (D-6) bulldozers have a purchase price of approximately $135,000. Bulldozers larger than the D-6 size are generally not appropriate for site work. Average costs given in Appendix F vary according to manufacturer and type of blade.

Trencher

This piece of equipment is commonly referred to as a "Ditch Witch." A variety of hydraulically powered tools and accessories are available for specific site applications.

Performance--
Trenchers are suitable for a variety of excavation/grading applications on site. They are generally wheeled and able to perform many of the same functions as larger backhoes, excavators, and bulldozers, but with a smaller carrying capacity. A variety of accessories are available, including:

- Backhoe
- Backfill blade
- Augers
- Remote handling

Application for Hazardous Waste Site Work--
Trenchers are readily available and easily transported. This piece of equipment combines the functions of several pieces of equipment into one unit for smaller sites. Applications for the use of trenchers include:

- Soil and sludge excavation
- Excavation of trenches for laying of pipe/lines
- Mixing of contaminated material with stabilization/solidification agents
- Site preparation
- Feeding material into treatment process

Cost--
Cost summary information for trenchers is not available.

Skid-Steer Loader

This piece of equipment is commonly referred to as a "Bobcat." It can be equipped with a variety of hydraulically controlled buckets, grapplers, and lifting attachments.

Performance--
Skid-steer loaders have excellent application for hazardous waste site work. They can perform many of the same functions as the much larger front-end loaders, and they can work in a much smaller area. Skid-steer loaders are ideally suited for indoor work, where they can operate with propane rather than diesel fuel. Although skid-steer loaders can traverse rough muddy terrain, they function best on relatively flat surfaces. A disadvantage of the Bobcat is its smaller carrying capacity. Figure 7 shows a typical skid-steer loader being used for site work.

Application for Hazardous Waste Site Work--
Skid-steer loaders are readily available and are easily transportable. Possible uses include:

- Drum loading and transport.
- Offloading of equipment from trucks.
- Soil and sludge excavation (fitted with shovel).
- Mixing of contaminated material with stabilization/solidification agents.
- Site preparation.
- Feeding material into treatment processes.

Cost--
Cost summary information for skid-steer loaders is presented in Appendix F. Purchase cost varies according to bucket width [$5200 (35-in. bucket) to $22,000 (63-in. bucket)].

Forklift Truck

Heavy-duty rubber-tired forklift trucks are available in capacities ranging from 1 to 50 tons; the most commonly used ones are 1-, 1.5-, and

Materials-Handling Equipment and Procedures 31

Source: PEI Associates, Inc. 1990

Figure 7. Skid-steer loader with hoe attachment.

2-ton vehicles. As shown in Figure 8, several accessories are available for specific loads. These include high-lift masts, handling attachments for cylindrical objects (drums), carton clamps, and fork side-to-side shifting mechanisms (Perry 1984).

Performance--
Forklift trucks are suitable for loading, offloading, and staging of equipment and drums at hazardous waste sites. The forklift truck has the advantage of being compact, maneuverable, and versatile (Wagner, et al. 1986). Although this piece of equipment requires a flat, relatively stable terrain, additional tractability can be provided with 4-wheel-drive units.

Application for Hazardous Waste Site Work--
Forklift trucks are readily available and easily transportable. General applications include:

- Offloading of equipment from trucks.
- Onsite hauling over short distances.
- Loading, staging, and transport of drums.

Cost--
Cost summary information for forklift trucks is presented in Appendix F. The purchase price of these trucks varies from approximately $30,000 to $85,000, depending on lift capacity, horsepower, and 2-wheel vs. 4-wheel drive.

DREDGING

Many hazardous waste sites require remedial techniques that entail removal and/or containment of contaminated sediments. The process of removing bottom sediments from bodies of water is referred to as dredging (U.S. EPA 1985a). Dredging is used for unlined surface impoundments containing surface liquids that cannot be removed, but contain sediments (sludge) that require removal. Dredging categories discussed in this section are mechanical, hydraulic, and pneumatic. The EPA document "Remedial Action at Hazardous Waste Sites" (U.S. EPA 1985a) contains a detailed description of dredging equipment and operations. Figure 9 shows an overview of several mechanical and hydraulic dredging operations. Table 9 presents a comparison of dredging equipment that is suitable for hazardous waste site work.

Mechanical Dredging

Capabilities--
Mechanical dredging techniques are frequently used in conjunction with standard excavation equipment (backhoes, draglines, clamshells, etc.). These techniques are most applicable to relatively shallow, low-velocity streams and rivers. Clamshell dredges, dragline dredges, backhoes, and bucket-ladder dredges are most commonly used for mechanical dredging. Mechanical dredging is generally applied in bodies of water less than 100 ft deep and having stream flows of 2 ft/s or less (U.S. EPA 1985a).

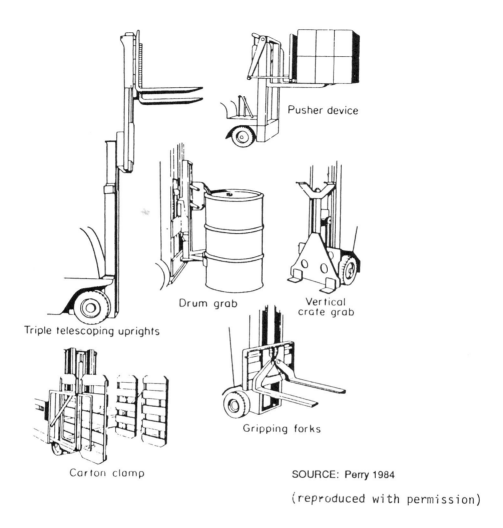

Figure 8. Various types of fork-truck attachments.

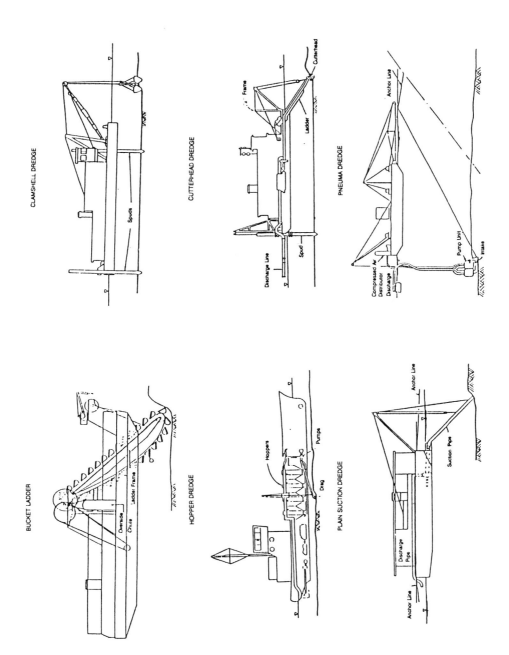

Figure 9. Mechanical and hydraulic dredging operations.
Adapted from: Hand 1978

TABLE 9. COMPARISON OF DREDGE EQUIPMENT THAT IS APPLICABLE AT HAZARDOUS WASTE SITES[a]

Factor	Mechanical			Hydraulic					Pneumatic		
	Dragline	Backhoe	Clam-shell	Plain suction	Cutter-head	Dustpan	Hopper	Port-able	Air lift	Pneu-matic	Oozer
Vessel draft, ft	[b]	[b]	[b]	5-6	3-14	5-14	12-31	1 1/2	3-6	NA[c]	7
Hinderance to traffic	Small	Small	Small	Significant	Some	Significant	Small	Some	Some	Some	Significant
Maximum wave height, ft	<3	<3	<3	<3	<3	<3	<3	<3	<3	<3	<3
Operation near structures?	Yes	Yes	Yes	No	No	No	No	No	No	No	No
Operation in open water?	Limited	Limited	Limited	Limited	Limited	No	Yes	No	No	Limited	Yes
Capable of consolidating sediments?	Yes	Yes	Yes	No	Yes	Yes	No	No	No	No	Yes
Susceptible to debris damage?	No	No	No	Yes	Yes	Yes	Yes	Yes	Yes	Yes	Yes
Suitable for liquid or solid removal?	Solid	Solid	Solid	Both	Both	Both	Both	Both	Both	Both	Both

[a] Source: U.S. EPA 1985a.
[b] Depends on draft of supporting structure. Most barges have a draft of 5 to 6 ft.
[c] NA = Not applicable.

Performance--
Clamshell (grapple) dredges are crane-operated and mounted on flat-bottomed barges or crawler tractors. The working depth of the clamshell, while theoretically only limited by the length of the cable, is realistically about 100 ft. Bucket capacities range from 1 to 12 yd^3. Because clamshell dredges leak heavily the Japanese have recently developed a tongue-and-groove clamshell that is watertight (U.S. ECE 1986).

The dragline dredge (Figure 10) can also be mounted on flat-bottomed barges or crawler tractors. Like the clamshell, the dragline can be used for any type of material. The drag cable pulls the dragline bucket through the material to be excavated. This provides a longer reach than the clamshell.

Bucket ladder dredges (Figure 9) use an inclined submersible ladder supporting a continuous chain of buckets that rotate around pivots at both ends of the ladder (U.S. EPA 1985a). As the buckets rotate around the bottom of the ladder, they scoop up the sediment, which is transported to and dumped in a storage area.

Backhoes also may be used for dredging purposes; these backhoes may be mounted on a barge. Because they have a more limited lateral and vertical reach than do clamshells or draglines, backhoes are infrequently used to remove contaminated sediment.

Application for Hazardous Waste Site Work--
Clamshell and dragline dredges are readily available and are capable of excavating materials at nearly in situ densities (U.S. EPA 1985a). They also generate large amounts of sediment resuspension. Silt curtains can be used to reduce sediment suspension. Clamshells are effective for deep-water excavation, whereas draglines are limited to shallow-water. Although clamshells and draglines have excellent lifting power, they are limited in mobility and rotation speed, which slows excavation operations.

Although backhoes are readily available for hazardous waste site work, they are limited by their reach. Smaller than clamshells or draglines, backhoes have greater mobility and manueverability than other mechanical dredgers.

Cost--
Extensive unit cost information is presented in EPA's handbook "Remedial Action at Hazardous Waste Sites" (U.S. EPA 1985a). Backhoe costs are also presented in Appendix F. Costs of crawler-mounted clamshells range from $95,000 to more than $1,000,000, depending on bucket capacity, horsepower, and manufacturer. Truck-mounted clamshell costs range from $60,000 to $650,000, depending on axle configuration, horsepower, and manufacturer. Renting/leasing information for hazardous waste site work is not available.

Costs of crawler-mounted draglines range from $100,000 to more than $1,000,000, depending on manufacturer and horsepower. Costs of truck-mounted draglines range from $55,000 to $500,000, depending on horsepower, axle configuration, and manufacturer. Renting/leasing information for site work is not available.

Materials-Handling Equipment and Procedures 37

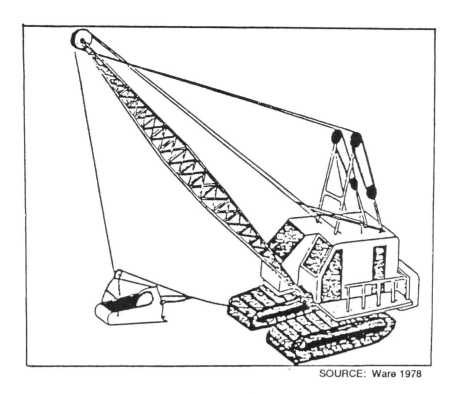
SOURCE: Ware 1978

Figure 10. Typical dragline.

Hydraulic Dredging

Capabilities--
Hydraulic dredges are used to remove and transport sediment in a liquid slurry (10 to 20 percent solids by weight). They can be used in waters with an appreciable flow rate (U.S. EPA 1985a). One of the disadvantages of hydraulic dredging is that, given the high liquids content of the dredged material, large settling/dewatering areas must be available. Five types of hydraulic dredges are commonly used: plain suction, cutterhead, self-propelled hopper, portable (Mud Cat), and pneumatic.

Performance--
Plain suction dredges rely on a centrifugal pump to provide suction to capture and transport excavated slurry. These dredges are pulled along the bottom of the pond or impoundment, and the dredged material is discharged through a pipeline. These units are used primarily for digging soft free-flowing materials.

Cutterhead pipeline dredges are widely used to transport waterbound solids. They are similar to the plain suction dredge, except that a rotating cutter loosens the material, which is then sucked through the pump. These dredges can be used to pump all types of alluvial materials, as well as clay and other compacted deposits (Pit and Quarry 1976). Cutterhead dredges are classified by the diameter of the discharge pipeline and range in size from 4 in. to 36 in. discharges. Concentrations of suspended solids from dredging operations using cutterheads range from 200 mg/L to 300 mg/L near the cutterhead to a few mg/L 2000 feet from the dredge (Averett, Perry, and Torrey 1989). The cutterhead dredge is capable of removing sediment with relatively small amounts of resuspensions extending beyond the immediate vicinity of the dredge (Raymond 1984).

Hopper dredges are similar to the plain suction dredges, except that they are self-propelled and are usually larger, ocean-going ships. Sediment/material raised from the bottom-dragged suction heads is pumped into storage tanks in the ship. To discharge its contents, the ship proceeds to a deep-water dumping area, where its hopper doors are opened to discharge the material.

Pneumatic Dredges

Capabilities--
Pneumatic dredges remove a higher proportion of solids than coventional dredging operations (up to 60 percent solids have been sustained during actual dredging operations). A disadvantage of the pneumatic dredge is that the pull of the compressed air cylinders decreases in shallow water, so that the effective operating depth is greater than 7.5 feet (U.S. ACE 1986).

Portable hydraulic dredging systems may be equipped with any of the units. They are designed to be easily transportable. Figure 11 shows an example of a portable dredge. One major advantage of portable systems is a vessel depth, allowing them to work in water less than 5 ft.

Materials-Handling Equipment and Procedures 39

SOURCE: Crisafulli Pump Co. 1989

Figure 11. Portable hydraulic dredging system.

Pneumatic dredges are hydraulic dredges that use a compressed-air pump and hydrostatic pressure to draw sediments/material to the collection head (U.S. EPA 1985a). They can be operated in shallow or deep water, and they are easily dismantled and transported. The capacity of a large pneumatic system is 2600 yd^3/h (Averett, Perry, and Torrey 1989).

Application for Hazardous Waste Site Work--
Surface impoundments for which hydraulic and pneumatic dredging is applicable include holding ponds; settling ponds; aeration lagoons; sludge or slurry pits; dewatering basins; and general industrial, storage, treatment, and disposal ponds. These impoundments may be natural depressions, artificial excavations, or diked containment areas (JRB Associates 1982).

Plain suction dredges can handle large volumes of material, but the sediments require extensive dewatering. Cutterhead dredges are most efficient when cutting and directing sediment/bottom material toward the pump, but the action of the cutterhead results in turbulence and resuspension of the sediment. Although all of these dredging systems have hazardous waste site applications, the portable hydraulic dredges are most applicable for many isolated sites.

Cost--
Extensive unit cost information is presented in EPA's handbook "Remedial Action at Hazardous Waste Sites," 1985. Purchase costs for a portable hydraulic-cutter suction dredge range from $44,000 for an 8-in.-discharge 365-hp unit to $3,000,000 for a 24-in.-discharge 4400-hp unit. Renting/leasing information is not available. No cost information is available for pneumatic dredges.

PUMPING

Many hazardous waste sites require the handling and disposal of large volumes of liquids, including liquids with a high solids content or corrosive characteristics. Pump selection depends on various factors, including the following:

- ° Properties of the liquid to be handled
  - pH
  - Viscosity
  - Temperature
  - Vapor pressure
- ° Required flow
- ° Intake and discharge pressures
- ° Metering
- ° Solids content

In general, pumps can be classified into two types: positive displacement and centrifugal. Figure 12 shows a classification chart of these pump types. A third type, the submersible pump is also discussed in this section.

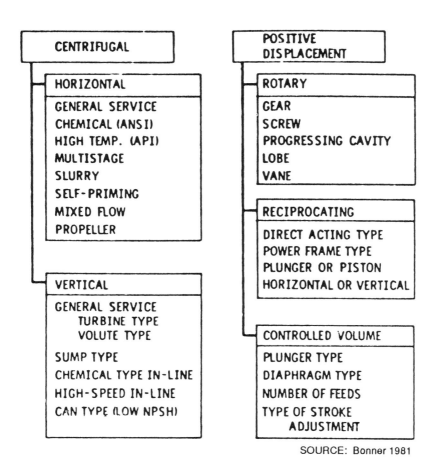

Figure 12. Pump classification chart.

Positive-displacement pumps may be either reciprocating (piston, plunger, or diaphragm) or rotary type. Positive-displacement pumps displace the liquid from the pump case by the reciprocating action of a piston or diaphragm or by the rotating action of a gear, cam, vane, or screw (Bonner et al. 1981). These pumps have a high overall operating efficiency, and they can deliver liquids at low velocities with high pressure. A disadvantage of positive-displacement pumps is that they are less efficient with low-viscosity fluid (internal slippage, air inclusion, etc.).

Centrifugal pumps (trash pumps) are the most commonly used pump in the chemical industry. Flow rates range from 2 to 10^5 gallons per minute (gpm). The advantages of centrifugal pumps are their simplicity, low initial cost, uniform (nonpulsating) flow, small floor space requirements, low maintenance requirements, quiet operation, and adaptability (Perry 1984). Centrifugal pumps convert velocity pressure generated by centrifugal force to static pressure. An impeller rotating at high speeds imparts velocity to the fluids. For field use, internal combustion engines (gasoline or diesel fuel) are used to run the pumps.

Submersible pumps can be small centrifugal pumps or piston (air operated) pumps and are used to drain shallow pits or sumps. These pumps are operational only when completely submerged in the liquid. Some types can operate with as little as 3/16 inch of fluid and can pump semisolids (U.S. EPA 1985a).

Application for Hazardous Waste Site Work

General pump selection depends on the required rate of flow and the physical/chemical characteristics of the material to be pumped. For hazardous waste work, several different materials (e.g., acids, dyes, organic solvents, brines, and caustics) may need to be pumped. A variety of liner materials have been developed to provide resistance to hazardous substances. These include Viton, Nordel, neoprene, Teflon, polypropylene, polyethylene, and natural rubber. Abrasive materials (e.g., titanium dioxide, lead oxide, and machine coolants) also may present a problem during pumping operations. Rubber and ceramic liners are effective for resisting abrasive wear (Perry 1984). Piston pumps should not be used to pump fluids containing abrasives, as leaks could develop in the pump packing seal.

Both centrifugal and positive-displacement pumps can be used with fluids having a high solids content. Reciprocating pumps have the advantage of being able to pump sludges. Screens can be used to remove oversized materials that might damage the pump.

Whereas none of the industries contacted mentioned specific uses of pumps or pumping systems, the chemical industry is known to have developed a wide variety of pumps for dealing with caustics, ceramics, abrasives, alcohol, acids, brines, paper manufacturing effluent, mining waste, sludges, solvents, resins, glue, and food. Applications for hazardous waste site work can be found among many of these pumps used by industry.

Cost

Renting/leasing information for various pumps is presented in Appendix F. Typical purchase costs are also given; however, costs vary widely, depending on horsepower, port size, interior lining, and manufacturer.

SIZE AND VOLUME REDUCTION

Size Reduction

Various kinds of size-reduction equipment are available. Equipment used to obtain uniform size reduction includes small grinders, chippers, roll crushers, jaw crushers, large grinders, shredders, rasp mills, hammer mills, and hydropulpers. The goal of size reduction is to obtain a final product that is both smaller in size and homogeneous. For general hazardous waste site work, size reduction does one of the following: 1) facilitates staging, storing, and ultimate disposal of debris/material; or 2) serves as one of the initial steps in feedstock preparation for treatment (i.e., attaining optimum size of debris or materials that can be handled by a given treatment technology).

Several sources (Tamm, Cowles, and Beers 1988; Tchobanoglous, Theisen, and Eliassen 1977) have considered the factors that affect the selection of size-reduction equipment for hazardous waste site work. These include:

- Properties and characteristics (density, moisture content) of material to be shredded (both pre- and post-shredding).

- Size requirements of the shredded material for treatment or containment.

- Availability of equipment (including mobility).

- Type of infeed system required to ensure efficient flow rate [infeed opening, batch vs. meter (regulated) feed methods].

- The ability of the treatment process to accept metallic materials and metallic debris commonly found on hazardous waste sites.

- Type of operation (continuous vs. intermittent).

- Site considerations (size of work area, terrain, weather conditions, access, and noise).

- Precautions taken for release of toxic materials or explosives within the size-reduction equipment.

- Ease of cleaning and decontamination of shredding equipment (including feed equipment).

Size-reduction equipment used for hazardous waste work was initially designed and used to handle bulk materials such as sand and gravel, topsoil, coal, iron ore, and crushed rock. Several companies (e.g., Powerscreen of America, Inc.; Shredding Systems, Inc.) have developed systems for the handling or preprocessing of hazardous waste. Shredding equipment can facilitate several treatment technologies, including the following (Powerscreen of America, Inc. 1988):

- Incineration--requires consistent feed rate and a predetermined particle size.

- Biochemical treatment--is augmented by having small-particle-size and high-surface-area raw material.

- Solidification/stabilization processes--benefit from uniform particle size; in some cases, the shredding equipment's receiving hopper can be used for mixing.

The shredder is the most commonly used piece of size-reduction equipment. Figure 13 presents an example of a self-contained portable shredding system for hazardous waste work. The most common type of shredder used in the municipal solid waste field is the hammermill, which consists of a central rotor with a hammer that causes a crushing action against the breaker plates. Other types of shredders include ring mills, grinders, flail mills, and ball mills.

Pug mills are also available for size reduction and homogenization (see page 198).

Shredders available for hazardous waste site work may be hydraulically, electrically, or diesel driven. They may be used strictly for shredding purposes or fitted with units that provide material separation capabilities. Several case studies were presented at the National Technology Seminar (1988) that illustrated actual onsite use of shredding systems. Shredding equipment has been used in conjunction with backhoes, front-end loaders, farm disks, forklifts, and grapplers to scalp and screen out such debris/material as batteries, tires, sheet metal, rock, clay, concrete pipe, paint cans, and railroad ties.

Volume Reduction

Because typical size-reduction equipment often results in significant volume reduction of the debris/material handled, several types of densifying equipment used in the municipal solid waste industry may be applicable at hazardous waste sites. Densifiers are primarily used at the end of the process line in MSW-to-energy systems to enhance the storability or transportability of the waste. Densifiers can be classified into five different pieces of equipment: briquetters, cubetters, extruders, pelletizers, and compactors. A report by Bendersky et al. (1980) contains a detailed description of densifying equipment used for MSW-to-energy systems. (Drum compaction is discussed in the Drum Handling and Removal subsection.)

SOURCE: Shredding Systems Inc. 1988

Figure 13. Self-contained portable shredder.

SEPARATION AND DEWATERING

Component Separation

The primary function of separation processes is to obtain two or more distinct waste streams separated on the basis of a specific characteristic such as size, density, or material type. Separation/classification of materials may be accomplished by either mechanical or manual means. Material separation at hazardous waste sites is important for several reasons. First, the screening of materials allows for more efficient operation of the chosen treatment process (i.e., screening for maximum size of debris/materials for a given technology). Debris/materials that exceed the maximum allowable size must be separated and treated/disposed of separately. Second, separation facilitates the staging and storing of debris/materials found on site by grouping like-sized material. Finally, recent evidence suggests that many contaminants will preferentially adsorb to fine-grained materials such as clay and organic material (U.S. EPA 1985a). Therefore, separation by grain size should reduce the volume of contaminated material to be treated.

Screening equipment can be divided into the following five general categories (Perry 1984):

- Grizzly screens--Sets of parallel bars set at predetermined spaces (can be stationary or vibrating)

- Revolving screens (trommel screens)--A revolving cylindrical frame surrounded by wire cloth--open at both ends

- Shaking screens--A rectangular frame lined with wire cloth (often used in conjunction with conveying system)

- Vibrating screens--Used for high capacity and efficiency (may be mechanically or electrically powered)

- Oscillating screens--Characterized by low-speed oscillation (often used with silk cloth)

Grizzlies are used primarily for scalping, i.e., removing a small amount of oversized material from material that is primarily fines (U.S. EPA 1985a). Moving screens (i.e., vibrating, shaking, revolving, and oscillating) are used to separate particles by grain size, typically in the size range of 0.125-in. to 6-in.

Two factors have a direct bearing on the selection of screen equipment. Overall width relates to capacity (i.e., the greater the width, the greater the capacity), and the length of the screen relates to efficiency (i.e., the longer the screen, the longer the residence time of the material) (Perry 1984).

Additional separation systems that may have application for hazardous waste site work include air separation (light vs. heavy components), flotation, and magnetic separation (ferrous vs. nonferrous metals). These systems have been discussed at length for MSW-to-energy operations (Runyon 1985; Savage and Shiflett 1980).

Selection of a separation system for hazardous waste work involves the following factors:

- Availability of equipment (mobility)

- Type of debris/material to be handled (physical/chemical characteristics)

- Volume of material to be handled

- Site considerations (size of working area, terrain, weather conditions, access, and noise)

- Ease of cleaning and decontamination of separating equipment

- Whether separation is to be accomplished dry or wet

- Type of feed-delivery system, including feed rate

From discussions with response personnel and OSCs, most debris/material separation is accomplished by hand-sorting. Hand-sorting is used primarily for the sorting of larger, bulkier materials in the initial staging of onsite materials. Several hazardous waste site descriptions have mentioned the use of screening devices such as grizzlies and vibrating screens in conjunction with shredders (National Technology Seminar 1988). The screening of materials enabled more efficient operation of the chosen treatment processes (incineration of PCB-contaminated transformers and capacitors and microbial degradation of creosote-contaminated soil). A detailed description of feed-stock preparation and handling is presented in a recently completed EPA report (Tamm, Cowles, and Beers 1988).

Dewatering

Dewatering is the removal or reduction of moisture from sludges or slurries. Dewatering techniques are used on hazardous waste sites where large volumes of sludge or sediment must be handled or treated. The ability to dewater sludge is important because it determines the volume of waste to be handled. If a sludge is dewatered to 5 percent solids, 4/5 of the volume is eliminated; if dewatered to 10 percent solids, a 9/10 volume reduction occurs (Robinson 1986). The contaminated water resulting from the dewatering will require additional treatment. A detailed discussion of dewatering techniques is presented in "Remedial Action at Waste Disposal Sites" (U.S. EPA 1985a). Several dewatering methods are briefly described here. These include the use of thickening and conditioning agents, centrifuges, vacuum filters, pressure filters, and dewatering lagoons.

Thickening Agents--

Sludges or slurries are generally sent to a thickener before going to dewatering equipment. Gravity thickeners and flotation thickeners are commonly used. Solids concentrations of 2 to 15 percent are achievable by the use of thickening agents (U.S. EPA 1985a). Thickeners aid in water removal and blending. It should be noted that the addition of thickening agents will result in an overall volume increase of the sludge or slurry, but is often necessary in order to transport and dispose of the waste.

Conditioning Agents--

Three chemicals (lime, ferric chloride, and polymers) are most often used before dewatering, regardless of the dewatering equipment used. These chemicals can be used separately or together to enhance settling (Robinson 1986).

Centrifuges--

The three types of centrifuges that have been developed are the disc, basket, and scroll centrifuges. Centrifuges operate on the basis of density difference separation, and are therefore highly applicable for sludge dewatering. They are applicable for dewatering soils and sediments ranging from fine gravel down to silt (U.S. EPA 1985a). Because centrifuges are compact, they require relatively little space and can be quickly mobilized and demobilized. Solids concentration ranges from 9 to 25 percent, and a solids capture of 85 to 97 percent is possible (U.S. EPA 1985a).

Vacuum Filters--

The most common vacuum filters consist of a cylindrical rotating drum that is submerged in the sludge. Solids concentration ranges from 20 to 40 percent, and a solids capture of 85 to 95 percent is possible; however, this process may require the use of large quantities of conditioning agents. An additional limitation is that incoming feed must have a solids content of at least 3 percent. Other types of vacuum filters are also available (e.g., belt filters).

Pressure Filters--

Pressure filters are the most powerful dewatering devices in use. Solids concentration ranges from 30 to 45 percent, and a solids capture of 95 to 100 percent is possible. One type of pressure filter consists of vertical plates held rigidly in a frame and pressed together by a large screw jack or hydraulic cylinder (U.S. EPA 1985a). Other types of pressure filters are also available (e.g., pressure leaf filters). As with the vacuum filter, conditioning agents are generally required. Both pressure and vacuum filters are applicable for mobile on-site work.

Dewatering Lagoons--

A dewatering lagoon is lined with clay, and a synthetic liner is used in conjunction with a gravity- or vacuum-assisted underdrainage system (U.S. EPA 1985a). Although dewatering lagoons are highly effective (a solids concentration of 35 to 40 percent and 99 percent solids removal), this technique is generally applicable to very large-scale operations, and it requires large areas and long setup times. Gravity filtration can also be used for smaller applications (e.g., sand filtration units).

CONVEYING SYSTEMS

Conveying systems are often used at hazardous waste sites, generally in conjunction with shredding and screening pretreatment of feedstock for onsite treatment (Figure 14). Several mobile conveying systems are available, including screw conveyors, belt conveyors, bucket elevators, vibrating or oscillating conveyors, and continuous-flow conveyors. A comprehensive description of these various pieces of equipment is found in the Chemical Engineers Handbook by Perry (1984).

The following factors should be assessed before selecting a conveying system for hazardous waste work:

- Capacity requirements
- Length of travel
- Physical/chemical characteristics of transported material
- Availability of equipment (mobility)
- Ease of cleaning and decontamination of conveying systems

STORAGE CONTAINERS, BULKING TANKS, AND CONTAINMENT

Storage Containers and Bulking Tanks

An often overlooked aspect of materials handling on hazardous waste sites is the selection of a storage system. Storage containers may range from small 5-gallon pails to 12,000-gallon or larger bulking tanks. These storage containers may be used simply to stage the material, or they may be used in the treatment process. The most commonly used storage containers are 55-gallon drums and 85-gallon overpacks (generally used for offsite disposal). Storage containers or bulking tanks may be constructed of polyethylene, polypropylene, or stainless steel. Care must be taken to determine the chemical characteristics of the waste to be stored so the construction materials of the disposal container chosen will not be incompatible with the stored material.

Hazardous waste container bag liners are also available for lining roll-offs, dump trailers, compactors, or any vehicle used to transport hazardous waste. These liners generally provide a triple-thickness of leak protection and are constructed of high-density polyethylene.

Fiberboard drums and boxes also can be used to store solids. These containers are generally used when the treatment option is incineration and both the container and the waste are to be incinerated. These drums can also be lined to provide limited liquid storage. (This is not approved by the Department of Transportation, however, so such containers cannot be shipped.)

A variety of collapsible or flexible containers are also available for hazardous waste applications. Collapsible rubberized containers that lie flat when empty and assume pillow configuration when full are available for liquids. Collapsible, self-emptying containers are also available for asbestos removal.

SOURCE: PEI Associates, Inc. 1990

Figure 14. Conveyor being used during treatment process for contaminated soil.

A wide selection of polyethylene tanks for storing hazardous waste is commercially available. Types of tanks include vertical, horizontal, and cone-bottom. Department of Transportation (DOT)-exempt tanks, transportation/storage tanks, and custom-molded tanks are also available with capacities ranging from 55 to 12,000 gallons. These storage tanks are constructed of high-density cross-linked polyethylene to enable them to retain materials that have a high specific gravity, withstand the corrosive effects of most acidic and alkaline solutions, and resist stress cracking. Typical chemicals that the tanks can handle are sulfuric acid, sodium hydroxide, hydrofluoric acid, sodium hypochlorite, phosphoric acid, and ferric chloride. The DOT-exempted tanks are capable of storing up to 50 percent hydrogen peroxide. Polyurethane-insulated polyethylene tanks and polyethylene-insulated steel and alloy tanks are also available. Cross-linked polyethylene imposes two limits: 1) the working temperature must not exceed 150°F; and 2) repairs must be made by patching because polyethylene tanks are not weldable or fusible.

Containment Systems

A variety of products are available for containment of spills of hazardous material. Sorbent booms, pads, sweeps, blankets, pillows, and particulates are available for containment of industrial chemicals, oils, and most hazardous waste spills. Sorbents are also available to separate oil from water on hazardous waste sites. Oil-skimming/absorbent systems are also available for oil/water separation. Vacuum trucks and units also fall under the heading of containment systems.

Vacuum Systems--

Industrial vacuum loaders are frequently used on hazardous waste sites. Sizes range from small 5-gallon wet/dry vacuums to large 5000-gallon units mounted on trucks. Skid-mounted units are also available; the most commonly used size is 1500 gallons.

Vacuum loaders are capable of handling liquids, high-solids-content sludges, or solids. Whereas smaller vacuum units have a wide application for most sites, vacuum trucks (because of their high carrying capacities) are generally used only at sites where more than 1500 gallons of material must be handled. Another aspect to be considered is the compatibility of the contaminant with the construction materials of the vacuum unit. Vacuum cylinders are generally constructed of carbon steel, stainless steel, aluminum, or nickel alloys; however, they also can be treated with a variety of coatings, including epoxy, fiberglass, and neoprene rubber (U.S. EPA 1985a). Appendix F contains information on renting and leasing of vacuum units and trucks. Purchase price varies according to capacity and manufacturer.

COMPACTION

Soil Stabilizer

A soil stabilizer is a soil additive that dries, strengthens, and bonds soil into usable backfill. Soil stabilizers minimize subsidence from restored excavation by drying the soil, and they increase compressive strength of the soil 3 to 16 times while retaining its original size and shape by weaving soil particles together. These soil additives can be used on a wide variety of soils to turn the spoil into reusable backfill, which eliminates the need to haul in replacement backfill. Soil stabilizers are not toxic to plant or animal life (plants can grow in soil that has been stabilized). Performance testing has shown that the stabilizers do not contribute to the corrosion of metal, nor do they attack plastic piping. The stabilized soil remains slightly permeable and does not act as a moisture barrier. It can also be reexcavated later.

Uses of stabilized soil include backfill stabilizer, access road base stabilizer (not a permanent wearing surface), surface erosion control, pole and marker stabilizer, and earthen berm construction.

Equipment

A variety of compaction equipment is available for site work, including sheepsfoot rollers, impact hammers, and rollers. These pieces of equipment can be hand-held or attached to backhoes, excavators, or skid-steer loaders. Heavy equipment with tracks (bulldozers, etc.) also may be used for compaction purposes.

MISCELLANEOUS EQUIPMENT AND PROCEDURES

Various smaller pieces of equipment are used to perform work at hazardous waste sites. Hand tools such as shovels, rakes, hoes, brooms, etc., are used in support of larger pieces of equipment. Chain saws and cutting torches may be used to cut wood/vegetation or metal tanks and drums. Air compressors may be used at sites that require the use of an air hammer to break up solid pieces of material (concrete, asphalt). Many sites have no electrical source, either because of disrepair or their remote location. In this case, a gasoline- or diesel-powered generator may be required as a source of electrical power.

Another piece of equipment frequently used for hazardous waste work is the lowboy truck. This vehicle is generally used for loading and transporting heavy pieces of equipment (bulldozers, loaders, etc.). Although lowboys are not directly used for site remediation, their cost and the available space on site must be considered.

Portable buildings have applications for hazardous waste sites where it is necessary to contain air emissions. These buildings are designed and prefabricated for rapid erection and redeployment by unskilled labor. The modular design allows the building to be extended or subdivided easily as the user's needs change. In addition, a crane can be used to lift the building in its finished form. The buildings are constructed of galvanized steel tube arch frames (no internal columns) covered with a PVC-coated polyester fabric approximately 0.03 in. thick. Various standard sizes are available, and customized buildings are available from manufacturers upon request. These structures are made to withstand high winds and extreme weather conditions such as blizzards and tropical heat. The buildings can be constructed with a double layer of skin for increased insulation, and they are easily dehumidified.

DRUM HANDLING AND REMOVAL

The handling and removal of drums on hazardous waste sites involve a variety of procedures, including excavation, hauling, loading, lifting, pumping, crushing, and shredding. The following equipment is typically used for these procedures:

- Bulldozers
- Drum crushers/shredders
- Front-end loaders
- Backhoes
- Forklifts
- Cranes
- Drum grapplers (hydraulic)
- Drum punches
- Drum pumps
- Barrel carts
- Nonsparking tools

Backhoes are the most frequently used piece of heavy equipment for drum handling. They are used to excavate, to transport, and sometimes, to crush the drums. Backhoes also may be fitted with hydraulically operated drum grapplers.

Drum grapplers are hydraulic backhoe and excavator attachments designed for handling barrels and other similar cylindrical containers, including those for hazardous waste. The grappler is equipped with a large 360-degree rotating-turntable mechanism and 3/4-inch nonsparking neoprene lining. It is designed for easy attachment and removal and for reducing labor while increasing material handling speed and safety.

Cranes, forklifts, front-end loaders, and bulldozers also are often used at drum-handling sites, depending on the number of drums to be handled and the available working space. For sites requiring the disposal of less than 150 drums, a backhoe or front-end loader is commonly used to do the crushing; the bucket is used to flatten the individual drums. At sites containing more than 150 drums, a drum crusher is generally used. Although using a drum crusher is slower, it is more cost-effective for disposing of a large number of drums (Figure 15).

Nonsparking hand tools constructed of molybdenum, manganese-bronze, or aluminum alloys are frequently used around drum-handling and excavation operations, especially with potentially highly flammable or shock-sensitive materials. Drum carts, dollies, pumps, and punches are also commonly used at drum-handling sites. Wagner et al. (1986) presents a comprehensive outline of drum-handling techniques used on hazardous waste sites.

SOURCE: Piqua Engineering Inc. 1989

Figure 15. Drum crusher used for hazardous waste site work.

ASBESTOS REMEDIATION

Asbestos remediation projects present special equipment problems because of the necessity of preventing the release of asbestos fibers from the work area. Generally, two abatement techniques are used for asbestos remediation: the conventional containment approach and the glove-bag technique.

In the conventional containment approach, an enclosure is established around the inside perimeter of the work area within which the asbestos-containing material is to be removed. Plastic sheeting is placed on all walls and the floor. This sheeting is sealed around the ceiling to form an enclosure that essentially envelops the entire area. This method is generally used for large work areas (PEI Associates 1989).

With the glove-bag technique, a small enclosure is established around each pipe covered with asbestos-containing thermal insulation by placing the glove bag around the pipe and taping or strapping it across the top and sides to form an air-tight seal. Workers then place their hands in the gloves inside the bag to remove contaminated insulation without contaminating the area outside the glove bag. With both the conventional containment approach and the glove-bag approach, a negative air pressure must be established in the work area as a secondary means of containment (U.S. EPA 1985b).

The following is a list of the equipment and materials commonly used for asbestos remediation:

- Plastic sheeting
  - Minimum 4 mils thick for walls and stationary objects
  - Minimum 6 mils thick for floors
- Tape for sealing joints
- Impermeable containers (metal or fiber drums, 6-mil plastic bags)
- Glove bags
  - 7-mil clear polyethylene bag with attached tool pouch and entry port for insertion of wetting tube and/or HEPA-vacuum* hose nozzle
  - 10-mil clear polyvinyl chloride (PVC) bag with integral 10-mil-thick PVC gloves, elasticized valve/port, and tool pouch
- Airless sprayer
- Vacuums (with HEPA filters)
- Hand tools (scrapers, wire cutters, sprayers, sponges, shovels, flexible wire saws)

The document "Guidance for Controlling Asbestos-Containing Materials in Buildings" (U.S. EPA 1985b) is available for further information concerning asbestos remediation projects.

---

* A HEPA vacuum is a vacuum equipped with a high-efficiency particulate air filter.

EMISSION CONTROL

At most hazardous waste remediation sites, the dust and/or vapor emissions that result from the excavation, loading/unloading, or transport of soil, sludge, or sediment must be controlled.

The following 14 commercially available dust and vapor control methods have been identified (Todd et al. 1988):

1) Water--Added topically to increase the density and cohesion of soils, which reduces dust emissions

2) Water additives--Surfactants added to the water to increase penetration and retention time

3) Inorganics--Inorganic salts (e.g., calcium chloride) absorb and chemically bind moisture. Pozzolanic material (cement and lime) can also be used

4) Organics--Oils, bitumens, and vegetable gums bind with soils and, because they have a lower vapor pressure than water, are retained longer

5) Foams--Block vapor and dust escape routes

6) Air-supported membranes--Used to enclose excavation areas

7) Acid gas neutralization additives--Ferrous compounds react with and return sulfurous gases below the surface

8) In situ treatment--Technologies that treat volatile organic compounds without excavation (steam stripping, radio-frequency soil flushing, etc.)

9) Self-supporting enclosures--Enclosures that direct purged air to air pollution control devices

10) Vacuum tanks--Used to remove liquid and/or solids for reduction of dust emissions

11) Covers, mats, membranes--Provide a physical barrier on the soil (straw, wood chips, and sludges also used)

12) Wind screens--Used to reduce wind shear over soils

13) Seasonal planning--Avoids excavation during excessively dry weather

14) Silt curtains--Used to reduce resuspended material resulting from dredging operations

LOW-LEVEL RADIOACTIVE WASTE

Radioactive waste remediation presents special problems for materials handling because of the need to protect workers and the public from radiation. Generally, the same type of equipment found at hazardous waste sites (crane, loader, forklift, etc.) is used at radioactive waste sites. Equipment operators are not usually shielded because this can cause a more severe safety hazard (e.g., obstructing the operator's field of view). Therefore, remote materials handling is frequently used.

A mobile, remote-controlled, teleoperated, multipurpose robot can be used for remote materials handling. This is a basic mobile platform onto which a variety of arms/manipulators or payloads can be mounted. It has operational capabilities in hazardous materials detection, analysis, and handling as well as nuclear/radioactive waste decontamination. The robot is suited for both indoor and outdoor work with its remotely selectable wheel or track system. It can climb a 40-degree incline with a maximum tilt of 27-degrees on sand, muddy soil, and rocky terrain. It can also climb stairs and steep obstacles.

Portable shot-blast cleaning systems have been used at low-level radiation waste sites to strip contaminated paint off floor surfaces. These systems have a completely enclosed centrifugal blast wheel in the cleaning head. The wheel spins and metallic shot is propelled from the blades and blasts the floor surface. Both shot and contaminants are sucked into a separation system which recycles the blasting media for reuse and removes the contaminants to an attached dust collector. This machine can be used on concrete floors, parking decks, and road surfaces. This technique of airless shot blasting leaves surfaces dry and chemical-free and eliminates dust pollution. Portable shot-blast cleaning is also applicable for other hazardous compounds adhering to paint or concrete.

Several concentration technologies that reduce waste volume by concentrating the radioactive species from the original waste form are available for liquid low-level waste, wet-solids low-level waste, and dry-solids low-level waste. For liquid low-level waste, technologies include evaporation, distillation, crystallization, flocculation, precipitation, sedimentation, and centrifugation. For wet-solids low-level waste, sedimentation, centrifugation, dewatering, drying, and dehydration can be used. Possible choices for dry solids low-level waste are compaction, shredding, and baling (EG&G Idaho, Inc. 1984).

Commercially available decontamination equipment for dry-solids low-level waste includes mobile dry-cleaning systems, mobile electropolishing systems, mobile ultrasonic cleaners, portable high-pressure washers and glass bead spray systems, and portable steam cleaning and degreasing systems (EG&G Idaho, Inc. 1984).

Specialized containers must be used for the storage and transportation of low-level radioactive waste. There are two types of shipping containers for radioactive wastes: Type A quantity containers and Type B quantity containers. Type A quantity containers are based on limiting the amount of radioactive material to avoid excessive exposure to the general public should an accident occur (<200 mRems/h on accessible surface of container). Type A packages must meet DOT requirements outlined in 49 CFR Part 171 and regulations governing normal conditions of transport (Malloy 1984). Examples of Type A quantity containers are high-density polyethylene storage tanks.

Type B quantity containers, by definition, are those whose quantities exceed Type A quantity container limits. Type B radioactive materials require special casks to limit their release into the environment if a serious accident should occur. Type B packages must meet DOT requirements and regulations concerning normal conditions of transport and hypothetical accident conditions. In addition, the packages must be certified by the Nuclear Regulatory Commission (NRC) (10 CFR Part 71) (Malloy 1984). Examples of Type B containers are shielded shipping casks with lead shield equivalences ranging from 1.81 to 4.58 inches.

Low specific activity (LSA) radioactive materials are exempt from DOT Type A packaging requirements. (LSA materials are those in which the radioactivity is uniformly spread and excessive radiation exposure will not occur even if the materials were to be released in a moderate accident.) These materials must be packaged in strong, tight, industrial containers and transported in exclusive-use vehicles. The DOT exemption is applicable as long as the radiation levels on the container surface do not exceed 1 rem/h. Type A quantities also must not be exceeded. Otherwise, the packages must be certified by the NRC and meet the requirements for normal conditions of transport (Malloy 1984).

EQUIPMENT DECONTAMINATION

The most commonly mentioned method for decontaminating and/or cleaning equipment on a hazardous waste site is with a high-pressure/hot-water laser (Figure 16). Generally, a concrete decontamination pad is built and a sump is dug. The rinsate caught in the sump (which contains a 5-gallon drum) is treated, if necessary, and then disposed of. For smaller jobs, a layer of plastic lining is used instead of building a decontamination pad and sump. Other methods of decontamination mentioned by hazardous waste site personnel include vacuuming; applying solvent, acid-based foam or gel, or surfactants; and using scrub brushes with water, solvent, acid-based foam or gel, surfactants, or sand blasting and repainting.

Ease of equipment decontamination after use was mentioned as a factor in selecting a particular piece of equipment. Highly porous pieces (made of wood, concrete, plastic) are generally disposed of along with other contaminated solids. Some sites establish "clean" areas that can be accessed by incoming and outgoing equipment. This separates the contaminated equipment from the uncontaminated equipment and reduces decontamination time.

A comprehensive evaluation of equipment decontamination was presented by Esposito et al. (1985).

Materials-Handling Equipment and Procedures 59

SOURCE: PEI Associates, Inc. 1990

Figure 16. High-pressure water laser being used for onsite decontamination/cleaning of a front-end loader.

6. Foreign Contacts

Representatives of foreign environmental agencies, research groups, and remedial contract companies dealing with the cleanup of hazardous waste sites were contacted to obtain information regarding equipment and processes used for onsite materials handling. Countries contacted included France, Canada, Norway, the Netherlands, Germany, the United Kingdom (U.K.), and Denmark. Overall, 54 environmental agencies, response contractors, research groups, and vendors were contacted in these seven countries.

Information was obtained from the Warren Springs Laboratory (WSL), which is the U.K. government's principal environmental research laboratory. The following research areas are currently being investigated at WSL:

- Soil decontamination
    - Beach cleanup systems for oil
    - Processing and fractionation of solids
    - Slurries and materials with poor handling characteristics

- Materials recovery
    - Rotary screens
    - Air classification
    - Air tabling
    - Flat-bed screens
    - Shredders
    - Magnetic separators
    - Conveyors, screw feeders, hoppers

- Wet solids handling

- Dry solids handling

One innovative application of materials-handling equipment in the U.K. is the use of a scraper attached to a backhoe for beach cleaning of spilled crude oil (Figure 17). The scraper consists of sheets of rubber sandwiched between wooden boards and clamped to the bucket of a backhoe.

Warren Springs Lab has also developed a Beach Material Washer for cleaning oil from beach material (sand, pebbles). The process involves loading the contaminated material into mobile concrete mixers for mixing with kerosene. This preconditioned mixture is then discharged into a chute leading to a screening trammel. Beach material less than 25 mm in size passes to the

Foreign Contacts 61

SOURCE: Warren Springs Lab 1989

Figure 17. Backhoe with scraper attachment used for beach
cleanup of spilled oil.

feed tank of the washer. Oversized material (i.e., greater than 25 mm) is rejected after a limited wash (water) in the partially submerged trammel. Washed sand and small pebbles (less than 25 mm) leave the washer by a dewatering spiral screw system. After dewatering, the discharge is conveyed to a stockpile. The unit has been developed as a fully operational 20-tons-per-hour prototype and is expected to be scaled up to 50 tons per hour.

The Water Research Centre in the U.K. has published a directory of equipment used for the application of sewage sludge to agricultural land (Hall 1988). This directory contains information concerning the selection and availability of pumps, piping, sludge storage systems, soil injection systems, spreaders, and tankers. Although all of these pieces of equipment have application for hazardous waste sites, little work has been done in the U.K. on innovative uses of materials-handling equipment.

Harwell Laboratory in the U.K. have developed the Waste Management Information Bureau. This computer data base contains bibliographic details as well as abstracts and keywords for more than 57,000 documents on waste management, resource recovery, and associated topics. This data base is a potential source of information concerning materials-handling equipment and procedures in the U.K. and can be accessed (for a fee) through the Harwell Laboratory (Environmental Safety Centre), Oxfordshire, U.K.

Overall, companies and laboratories in the U.K. that specialize in the cleanup of hazardous waste sites have reported using many of the same types of equipment and procedures as those used in the United States. Typical pieces of equipment mentioned included:

- Front-end loader
- Bulldozers
- JCBs (backhoes)
- Trammel screen
- Centrifugal pump
- Positive displacement pump
- Shredder
- Conveyor

Hazardous waste sites in the U.K. are generally cleaned up either by excavation, removal, and landfill of contaminated material or by covering contaminated land with uncontaminated material (depending on the end use). No centralized program currently exists for the cleanup of hazardous waste sites in the U.K. As a result, available information concerning all aspects of hazardous waste site cleanup (including materials handling) in the U.K. is difficult to access.

Participants in the NATO/CCMS pilot-study meeting on remediation technologies for contaminated soil and groundwater held in The Netherlands in 1988 also provided lists of contacts that have dealt directly with onsite remediation in Germany. Several volumes of information were received concerning hazardous waste site remediation in Germany. All of the information received was written in German. Time constraints have prohibited any in-depth translation. Some of the German response personnel contacted stated that much of the equipment and procedures used for site remediation was "patent pending," and they were unable to provide details concerning materials handling.

7. Case Studies

The 22 case studies presented in this section illustrate specific applications and problems involving materials-handling equipment and procedures at different hazardous waste sites. The information included in these case studies was obtained from the following sources:

- RODs
- OSC reports and files
- ERCS files
- Communication with onsite response personnel
- Communication with OSCs for specific sites

In addition to the case studies presented in this section, Appendix E contains an overview of 67 sites (including the 22 case studies) from all 10 regions, including contaminants, debris/material handled, and primary debris handling procedures and equipment.

The specific information needed to write these case studies was obtained by visiting U.S. EPA Regional Offices. OSC reports and accompanying files were reviewed. Region IX was not visited because the needed information was physically unavailable as a result of the 1989 San Francisco earthquake. Availability and completeness of information varied considerably from Region to Region. Many files were unavailable because of litigation at the site. In several instances, the files and records for a particular site could not be located during the visit. Available files and reports usually contained incomplete information, regarding equipment usage.

An additional problem involved accessing confidential business information (CBI) at the EPA Regional Offices. Most Regions do not allow access to the CBI even when a Freedom of Information (FOI) request is filed. This presented an additional problem because most of the equipment information was presented with CBI cost information. The majority of the equipment usage information was obtained from CERCLA Daily Work Orders.

This section includes site history, contaminants, site description, equipment and problems encountered for 22 sites. Completeness of the information available varied for the various sites discussed, especially for the equipment used. An asterisk next to a piece of listed equipment indicates that the information as to the exact model is unknown.

WESTERN SAND AND GRAVEL, BURRILLVILLE, RHODE ISLAND (REGION I)

Site History

This semi-rural 12-acre site was used to dispose of septage and chemical waste. The wastes were dumped into unlined seepage lagoons and allowed to infiltrate into the soil and eventually to percolate to the groundwater. Available hazardous waste manifest records indicate that during the 1-year period 1978-79, 470,000 gallons of chemical waste was dumped at the facility. Records were not available for quantities of waste dumped before 1978.

Contaminants

The primary contaminants of concern included:

Chlorobenzene	1,1-Dichloroethane
Toluene	Trans-1,2-dichloroethylene
Ethylbenzene	Methylene chloride
Xylene	Trichloroethylene
Benzene	1,1,1-Trichloroethane

Site Description

Existing contamination at the site included:

- Approximately 400 yd^3 of sludge
- Contaminated liquids in the lagoon
- Contaminated groundwater

Equipment

Table 10 presents a detailed equipment list. Liquids from the lagoon were pumped out with skid vacuums equipped with basket strainers. Sludge was also pumped from the storage lagoon and solidified on site by shovel dozers. A safety dike was constructed around the solidified material for containment until it was packed in drums and dump trailers. A backhoe was used to crush 29 drums.

TABLE 10. EQUIPMENT USED AT THE WESTERN SAND AND GRAVEL SITE

Boat (without motor)	Generator
Dump truck	Electric wrench
Bobcat with front-end loader attachment	Basket strainers (2 in. by 1/8 in.)
2500-gal vacuum unit	Water laser (10,000-psi)
1000-gal skid vacuum unit	Diaphragm pump
25,000-gal vacuum rig	Drum pump
Shovel dozer (CAT 955)	Chemical-resistant hose
Shovel dozer (Case 350)	
Backhoe (Case 680)	

Problems Encountered

No major materials-handling problems were reported.

Final Disposition

Forty-two thousand gallons of liquid was pumped out of the storage lagoon and disposed of off-site. Two hundred ninety-six drums of material were disposed of after solidification with sawdust. Twenty-nine drums were disposed of after being crushed on site.

IRON HORSE PARK, BILLERICA, MASSACHUSETTS (REGION I)

Site History

This 15-acre site operated as a landfill from 1943 through 1975 for the disposal of asbestos waste. The landfill received asbestos sludge, dust, and waste board from an asbestos-insulating-board manufacturer. The threat posed by the site was from open uncovered piles of friable asbestos.

Contaminant

The major contaminant of concern was asbestos fibers.

Site Description

Exposed asbestos piles were randomly located throughout the landfill area. The area was vegetated and heavily wooded. Depth of the asbestos piles was unknown. Airborne fiber levels ranged from none detected to 0.006 fiber/cc. The work conducted at the site involved covering (capping) and stabilization of the asbestos piles. An estimated 6000 yd^3 of asbestos was located in the wooded area around the landfill.

Equipment

Asbestos contamination of the access road required that the road be covered with gravel and widened before truck traffic could be allowed onto the site. Spreading of the gravel and grading of the road were accomplished with a CAT D6 bulldozer and a Case 850 angle bulldozer. This site required extensive preparation prior to accomplishing the work. In addition to the road work, bulldozers and chain saws were used to clear extensive vegetation and trees. A detailed equipment list is presented in Table 11.

TABLE 11. EQUIPMENT USED AT THE IRON HORSE PARK SITE

Bulldozer (CAT D6)	Storage trailer (30 ft)
Bulldozer (CAT D4)	Lowboy
Bulldozer (Case 850)	Backhoe
OTR tractor	Front-end loader (CAT 955)
Rolloff box	Chain saw
Fire hose (1000 ft)	Generator

Problems Encountered

Extensive rainfall delayed site operations and the application of the cover material. These delays resulted in an extension of the period of performance. During drier days, daily dust control by the use of fire hoses was necessary to ensure that no asbestos fibers became airborne.

Final Disposition

A small number of bags filled with asbestos were transported from the site and disposed of.

INDUSTRIAL LATEX, WALLINGTON, NEW JERSEY (REGION II)

Site History

This 9.67-acre site, which contained two main buildings comprising 18,000 ft^2, was used to produce adhesives, resins, and other chemical products. Several fires had occurred in the buildings over the years.

Materials left on site included the following:

- 1300 drums and pails
- 2 above-ground fuel tanks
- 17 below-ground tanks
- 30 production vats
- 200 buried and partially buried drums

Contaminants

The following contaminants were found on site:

PCBs	Xylene
Benzene	Picco resins
Ethylbenzene	Hexane
Toluene	Methyl ethyl ketone
Nitrocellulose	Perchloroethylene
Barium	1,1-Dichloroethylene

Site Description

Many of the drums found at the site contained flammable and/or shock-sensitive substances and were either leaking or in a deteriorated condition. Several nearby wells had to be closed because of contamination from the site. The removal action performed at this site occurred during 1986. The following work was ordered to be performed:

- Fence construction and security
- Drum staging and segregation
- Removal of shock-sensitive, explosive, or highly flammable materials
- Removal of all material stored in drums, vats, and tanks

- Crushing and disposal of empty drums
- Cutting up and disposal of empty drums
- Debris removal on site (scrapwood, paper, etc.)

Equipment

A tracked backhoe with a drum grappler attachment was used to load the drums into the bucket of a backhoe for staging. A forklift with a drum grappler attachment was also used to stage the drums prior to shredding. Approximately 1000 drums were shredded over 5 days. A forklift and trackhoe were used to deposit the drums into the shredder. The shredded metal and PCB solids then fell into a box below the shredder teeth, where kiln dust was mixed as a solidifying agent. The mixture was then carried by a backhoe to a staging area.

Storage tanks were cut up for disposal using acetylene torches or a Cat 215 trackhoe with a shear attachment. Potentially explosive or flammable underground storage tanks were cut with a high-pressure water laser.

Aqueous wastes were pumped to a holding pool (6200-gallon capacity). Tanker trucks were used to remove a total of 15,000 gallons of aqueous waste.

Semi-solid wastes were also contained in two 20-yd^3 rolloff boxes. Liquids were drained off through valves attached to the rolloff box bottoms. A 125-ton crane was used to hoist the rolloff boxes onto wooden supports so they could be drained. The remaining flammable material was removed by a trackhoe with a bucket to scoop out the material, and it was then deposited within a bermed area for solidification. Shears were then attached to the trackhoe to cut up the rolloff boxes into 4-ft x 4-ft sections. Table 12 lists the equipment used.

TABLE 12. EQUIPMENT USED AT THE INDUSTRIAL LATEX SITE

Backhoe (CAT 215)	Pressure washer
Front-end loader (CAT 930)	Water laser
Skid-steer loader	6200-gal holding pool
Air compressor	Crane (125-ton)
Acetylene torch	Submersible pump
Diaphragm pump (2 in.)	Welding torch
Hand tools	12,000-gal holding pool
Backhoe (Case 580)	Barrel cart
Hydraulic shears	

Problems Encountered

Despite the major materials-handling effort involved, this site encountered few problems. The tank cutting by water laser was slow and expensive, but necessary to eliminate spark potential during cutting.

It was also necessary to control water runoff from the site to contain PCB migration offsite. An earthen berm was constructed to retain contaminated runoff, which was then channeled through one of three sand filters or septic tanks.

Final Disposition

Table 13 presents a breakdown of the amounts and disposition of wastes found at the Industrial Latex Site.

TABLE 13. SUMMARY OF OFF-SITE DISPOSAL OF WASTE FROM THE INDUSTRIAL LATEX SITE

Waste type	Amount, gallons	Disposition
Shock-sensitive/flammables	15	Detonated
Flammable organic liquids	2,940	Recycled
	1,441	Treated/Destroyed
Non-PCB aqueous	115,000	Treated/Destroyed
PCB aqueous	4,200	Incinerated
Flammable PCB solids	12,048	Incinerated
Nonflammable PCB solids	113,050	Landfilled

INTERNATIONAL METALLURGICAL SERVICES, NEWARK, NEW JERSEY (REGION II)

Site History

The 45,000-ft^2 site was an abandoned precious metals refining facility. A four-story structurally sound brick building still stands on the site. Material left on site included:

- Approximately 30 yd^3 of spent photographic film
- Piles of scrap metal, tanks, pipe, and miscellaneous debris
- More than 50 drums
- 450 laboratory containers
- 50 storage tanks
- Asbestos insulation

Contaminants

Contaminants found on site were as follows:

Asbestos	Chromic acid
Copper	Phenol
Lead	Cyanide
Mercury	Pyridine
Beryllium	Hydrazine hydrate
Silver	Nitric acid

Site Description

Many of the drums had either spilled contents or were leaking. The spent photographic film was in piles on the ground. Air samples taken in the building revealed the presence of asbestos and cyanide. The following work was ordered to be performed.

- Site stabilization
  - Debris removal
  - Removal of shock-sensitive material
  - Drum staging
- Recovery of salvageable/recyclable material
- Lab packing of known chemicals
- Consolidation of drummed waste
- Crushing of empty drums/lab containers
- Removal of spent film
- Removal of asbestos material around the pipelines

Equipment

The 55-gallon drums containing unknown materials were remotely opened by using a backhoe. One hundred three drums were crushed with a CAT 215 excavator bucket and then covered until loaded into rolloff boxes. A 10-ft x 10-ft crushing pad constructed of sand and lined with lime was constructed to crush 468 containers of known and unknown liquids and solids. Approximately 135 yd^3 of spent photographic film was loaded into 55-gallon fiber drums by using hand shovels, a vacuum cleaner, and a backhoe.

A major materials-handling effort at the site involved asbestos removal. Approximately 400 linear feet of exposed asbestos containing material (ACM) on pipes and fibrous material on the floor needed to be removed and disposed of. The following procedure was used to remove the asbestos:

1) The decontamination area was set up at the southwest stairwell.

2) A critical barrier of plastic sheeting was set up to enclose the stairwell.

3) Glove bags were used (6-mil transparent polyethylene with plastic arms).

4) The bags were installed to provide an enclosure of the section of pipe or fitting from which the ACM was to be removed. The sides of the bags were cut and placed around the pipe/fitting and the open edges folded and duct-taped to form a proper seal.

5) The painted canvas wrapping was cut away, and "amended" water was sprayed through a small incision in the bag.

6) As soon as the ACM was adequately wetted, it was removed from the pipe or fitting and placed in the glove bags.

7) After removal of the ACM, the pipe or fitting was thoroughly wire brushed and wet-wiped.

8) A HEPA vacuum machine was used to remove any remaining asbestos fibers prior to removal of the glove bags.

9) Glove bags containing the ACM waste were placed in polyethylene bags to achieve the necessary double bagging required for transportation to the approved disposal site.

10) Visual inspection of pipes and fittings was undertaken by the site manager. Recleaning of some sections of the pipes and fittings was ordered where traces of the ACM were detected.

11) The pipes and fittings were encapsulated after reinspection confirmed that the areas appeared sufficiently clean.

Table 14 lists the equipment used at this site.

TABLE 14. EQUIPMENT USED AT THE INTERNATIONAL METALLURGICAL SERVICES SITE

Excavator/backhoe (CAT 215)	Pallet lifter
Industrial vacuum cleaner	Hand tools (shovels)
Remote drum opener	Drum agitator
Forklift	Glove bags

Problems Encountered

Overall, few materials-handling problems were encountered. Constant clogging of the vacuum cleaners by the spent photographic film flakes required the use of shovels instead. The vacuum cleaners were used to skim the top layer of soil underneath the film piles.

Final Disposition

Table 15 presents a breakdown of the amounts and disposition of wastes found at the site.

TABLE 15. SUMMARY OF OFF-SITE DISPOSAL OF WASTE FROM THE INTERNATIONAL METALLURGICAL SERVICES SITE

Waste type	Amount	Disposition
Crushed drums	One 30-yd^3 rolloff box	Landfilled
Debris	Three 30-yd^3 rolloff box	Landfilled
Spent film	497 drums	Incinerated
Labpacks	105 drums	Incinerated
Asbestos	87 bags	Landfilled

BRUIN LAGOON NO. 2, BRUIN, PENNSYLVANIA (REGION III)

Site History

This 4-acre site consisted of a 2-acre open lagoon and a 2-acre closed lagoon. The site was a repository for process wastes from an on-site oil refinery. Concentrated sulfuric acid was used to treat the crude oil during the manufacturing process. The spent acid was deposited into the lagoon with other refinery wastes, such as bauxite, bone powder, carbon filter cakes, spent alkali, and coal fines. Estimates are that the lagoon contained two million gallons of acidic white oil sludge.

Contaminants

Contaminants found on-site were as follows:

Hydrogen sulfide gas   Organic acids
Sulfuric acid   Sulfur dioxide

Site Description

During the initial removal action, heavy equipment broke through what was thought to be the bedrock bottom of the lagoon. The bottom turned out to be a bauxite-like layer. The breakthrough resulted in the discharge of a 2-ft geyser of concentrated acid and acid mist. The fissure was covered, and boreholes were installed to monitor subsurface conditions. The initial removal activity also included the removal of hazardous materials in tanks. During the second removal action, the lagoon was filled in, the surface was stabilized, and 13 monitoring wells were installed to release trapped gases and to evaluate the conditions beneath the crust.

Equipment

Heavy earth-moving equipment (e.g., excavators and bulldozers) were the primary materials-handling equipment used on the site. Table 16 lists the equipment used.

TABLE 16. EQUIPMENT USED AT THE BRUIN LAGOON SITE

Rubber-tired loader (4.5 yd^3)*	Barrel cart
Bulldozer*	1.5-in. pressure pump
Backhoe*	1.5-in. suction and discharge hose
Hand tools	OTR tractor
Lowboy	

Problems Encountered

Unexpected delays and higher costs were encountered while a large bulldozer equipped with a ripper blade was located to break up the sludge that had solidified during the first removal action. During the backfilling operation of the open lagoon, upwelling sludge also became a problem and resulted in domed reservoirs of sludge. Reinforcment was added to the berm to ensure that the sludge would be contained within the site.

Final Disposition

No waste material was removed from this site.

AMBLER ASBESTOS TAILINGS PILE, AMBLER, PENNSYLVANIA (REGION III)

Site History

This 15-acre site is an active asbestos-processing facility located in a residential/industrial area. Two tailings piles of exposed asbestos-bearing material were located adjacent to homes and a playground. Wipe and bulk surface samples showed positive for asbestos and resulted in closure of the playground.

Contaminants

The primary contaminant was asbestos (chrysotile and amosite).

Site Description

The areas of concern were two asbestos piles covering 6.2 acres. The Locust Street pile was 1200 feet long, 300 feet wide, and 60 feet high. The plant pile was 900 feet long, 750 feet wide, and 60 feet high.

Equipment

Bulldozers, front-end loaders, and backhoes were used to cover the slopes with soil and to compact the soil covering the piles. Dust-control measures were implemented to cut down on airborne asbestos emissions. A drainage system was also installed to direct the runoff from the covered piles away from the city's storm drains. Table 17 lists the equipment used.

TABLE 17. EQUIPMENT USED AT THE AMBLER ASBESTOS TAILINGS PILE SITE

Backhoe*	Cutting torch
Barrel punch	OTR tractor
High-pressure washer	Lowboy
Hand tools	Bulldozer (CAT D3)
Portable air blower	Bulldozer (CAT D6)
Drum grappler (hydraulic)	Sheepsfoot roller
Front-end loader (4.5 yd^3)*	Erosion control matting
Forklift (2-ton)	

Problems Encountered

Because of the height and steep slope of the piles, the contractor experienced difficulties in covering and compacting them. The steeper slopes could not be cut to make them more gradual or terraced because of concerns for air releases of asbestos. Thus, the process of covering the slopes was very slow and difficult. A large bulldozer was used to winch a smaller bulldozer up and down the slope in some areas, but most areas required the use of a sheepsfoot roller. In most cases, hand tools were used to spread the soil on the slopes. The steepness of the slopes also presented an erosion problem that required the use of soil stabilization matting. Matting was unrolled over the slope and secured with metal staples or stakes.

Riprap also needed to be installed at the bottom of one portion of the Locust Street pile, where a creek had eroded the bottom portion of the pile.

Working space was also a problem at the site. Numerous delays resulted from the limited access and movement areas available to equipment drivers and operators.

Heavy rains during the operational period resulted in erosion and runoff problems. The muddy conditions on site resulted in significant time delays for vehicle decontamination by high-pressure washers. As many as 150 loads of soil per day were delivered to the site, and each truck had to be washed down before leaving the site. As a backup, a fire hose was used to remove mud tracked after the tire washing.

Final Disposition

No waste material was removed from the site, but roughly 600,000 yd^3 of asbestos-laden material was stabilized.

ABERDEEN PESTICIDE SITE, ABERDEEN, NORTH CAROLINA (REGION IV)

Site History

This site began operations as a formulator and manufacturer of pesticides. Pesticide and fertilizer residues were buried in pits and trenches on the site. (The site consists of three separate areas: the McIver Dump Site; the Twin Sites; and the Fairway Six Site, located on the Pit Golf Course.) The areas of concern are five separate areas contaminated with pesticide waste on the Fairway Site.

Contaminants

The contaminants of primary concern are:

Toxaphene
Endosulfan sulfate
Heptachlor
DDT
DDD
Lindane
Alpha-BHC
Beta-BHC
Delta-BHC

Site Description

The removal action involved the excavation of four trenches on the Pit Golf Course. The fairway that contained the trenches posed a threat to groundwater and surface water runoff. The removal action at this site involved the excavation of soil for eventual incineration.

Equipment

A tracked backhoe and front-end loader were used for the excavation of the soil. The soil was transported to a staging area, loaded into a hopper, and then moved by conveyor to a power screening apparatus. Discharge from the screening apparatus was stored on a 30-mil PVC liner. After all contaminated material was placed on the screen, a top cover liner was chemically affixed to the bottom liner. Table 18 lists the equipment used.

TABLE 18. EQUIPMENT USED AT THE ABERDEEN PESTICIDE SITE

Front-end loader*	Garden hose pump
Backhoe*	OTR Tractor
Vibrating screen	Lowboy (20- and 50-ton)
Conveyor	30-mil PVC liner
Loading hopper dump truck	Barrel shredder

Problems Encountered

Because of the large amount of material rejected by the power screening device, a shredder was installed on one of the screens. The shredder was used to attain a uniform size (<2 in.) of the rejected materials (e.g., crushed drums, bricks, large pieces of rubbish, and tires).

Final Disposition

Approximately 22,000 yd^3 of soil and an additional pile of crushed drums were temporarily stored on-site awaiting incineration.

A. L. TAYLOR SITE, BROOKS, KENTUCKY (REGION IV)

Site History

This 22-acre site was an uncontrolled industrial waste dump. It was used for the disposal of chemical liquids, sludges, and crushed drums. At one point, more than 17,000 drums were observed on the surface. The site had been used as a drum-cleaning and recycling operation, the drum contents had been disposed of in excavated pits on the property.

Contaminants

The primary contaminants of concern include:

Xylene
Methylene chloride
Phthalates
Toluene
Vinyl chloride
Aliphatic acids

Acetone
Methyl ethyl ketone
Barium
Zinc
Copper
Chromium

Site Description

Initial removal actions by six Responsible Parties in the early 1980's removed one-third of the waste from the site. A later EPA response removed 4000 drums, provided site grading, and directed surface water runoff to a retention pond. An on-site treatment facility was installed to treat the pond water before its discharge to a nearby creek.

An estimated 5,000 to 18,500 drums remained buried on the site. The following work was ordered performed on the site:

°   Removal of surface water from the site

°   Securing of pond sediments, sludge, and materials from low-lying area beneath the cap

°   Excavation of drum burial sites and removal of drums and other debris to a more secure area

°   Clearing, leveling, and compacting it of vegetation.

°   Installation of a cap cover to contain the waste material

°   Construction of a surface-water drainage diversion route to surface water around the cap area

Equipment

This site represented a major excavation effort. Trenches were dug to excavate the drums and debris, and a retention pond was dug to contain runoff water. Drum crushing operations were conducted with a backhoe and loaders. A detailed equipment list is provided in Table 19.

TABLE 19. EQUIPMENT USED AT THE A. L. TAYLOR SITE

Backhoe (CAT 205)
Dump truck
Electrical pump (3-in.)
Diaphragm pump (2-in.)
Trash pump (2-in.)
High-pressure washer
Discharge hose (3-in.)
Polyethylene holding tank
Excavator (CAT 225)

Air compressor
Generator (5- and 50-ton)
Lowboy (20- and 50-ton)
Bulldozer (CAT D3)
Bulldozer (CAT D6)
Bulldozer (CAT D65)
Farm tractor
Front-end loader (CAT 950)
Water trailer

Problems Encountered

The major materials-handling problem encountered was that the buried drums trapped water and wastes below ground, which resulted in a "spring" effect when excavation started. To contain this water/solvent, the southern end of an existing lagoon was dammed and the trench was lengthened to allow drainage into the dammed area. A retention pond was dug at the northern end of the trench to contain runoff in that direction. A granular activited carbon (GAC) unit was brought on site to treat this water/solvent mixture before discharge to a nearby creek.

Final Disposition

No waste was removed from the site. The crushed drums and debris were buried in a new excavated cell measuring 50 ft x 50 ft x 12 ft.

MIDWEST PLATING AND CHEMICAL CORP., LOGANSPORT, INDIANA (REGION V)

Site History

This site is an abandoned electroplating facility. The initial response resulted from the release of fumes through a broken window. The following materials were left on site:

- 16 abandoned plating lines
- Plating vats containing caustics and acids
- Storage tanks
- Approximately 100 drums containing unknown materials
- 5-gallon plastic containers

Contaminants

The following contaminants were found on site:

Tetrachloroethylene	Hydrogen cyanide gas
Chromic acid	Chromium
Potassium chromate	Cadmium
Organic peroxides	Hydrogen chloride

Site Description

The abandonment of this electroplating facility led to the release of hazardous vapors, including hydrogen cyanide gas and hydrogen chloride. The electroplating vats contained both highly acidic and alkaline liquids. Several underground chambers that had been used as sinks had collected large amounts of sludge and miscellaneous solid debris. Performance of the following work was ordered at this site:

- Removal and disposal of all hazardous material
- Decontamination of the building and its contents
- Sampling and analysis

Equipment

Table 20 lists the equipment required for cleanup of this site. The 16 plating lines had to be drained, dismantled, and decontaminated (by water laser). Miscellaneous debris had to be staged, decontaminated, and disposed of. Drums were staged inside by using a forklift and loader. Empty drums were crushed; because there were fewer than 100 drums, a backhoe was used for the crushing. The bulk of the materials-handling effort at this site involved sludge and solids removal from the underground chambers.

TABLE 20. EQUIPMENT USED AT THE MIDWEST PLATING AND CHEMICAL CORP. SITE

Box truck (2-ton)	Drum vacuum (wet/dry)
Stake-bed truck (2-ton)	Water laser (8000-psi)
Bobcat	Water laser (3000-psi)
Forklift	Holding pool (12,000-gal)
Diaphragm pump (2-in.)	Submersible mixer (2.5-hp)
Diaphragm pump (3-in.)	Crane (14-ton)
Electric submersible pump (2-in.)	Skid vacuum unit (1500-psi)
Front-end loader, wheeled (1.5-yd^3)	Backhoe (Case 580)
Air compressor (185-cfm)	Lowboy (20- to 40-ton)
Electric winch (1-ton load)	Chain saw
Vacuum tanker (1800-gal)	Generator (5-kW)

Problems Encountered

The major materials-handling problem encountered involved the removal of the plating sludge and miscellaneous debris (plastic piping, pails, empty drums) from the underground chambers. A vacuum tanker (or Supersucker) was able to pick up the sludge, but not the solid debris. This problem was overcome by using a downsized rolloff box structurally reinforced to handle an attached electrical winch for hoisting the sludge out of the chamber. A cut-down 55-gallon drum was lowered into the chamber, where it was manually loaded (by workers with shovels). The drum (2/3 full) was then hoisted up and dumped into the rolloff box, where it was mixed with kiln dust for solidification. A front-end loader (bucket) was used for mixing, and a Bobcat was used to remove the mixture for staging. This setup worked reasonably well, but both the winch and the electric motor had to be replaced.

Final Disposition

Table 21 presents a breakdown of the amounts and disposition of wastes found at the Midwest site.

TABLE 21. SUMMARY OF OFF-SITE DISPOSAL OF WASTE FROM
THE MIDWEST PLATING AND CHEMICAL CORP. SITE

Waste type	Amount	Disposition
Cyanide mixture	212 tons	Landfilled
Hazardous waste liquid	6 drums	Recycled/recovered
Plating waste	99,100 gal	Treated
Cyanide pretreatment (peroxide)	2 drums	Treated
Cyanide pretreatment	-	Treated
Decontamination water	5,000 gal	Treated
Waste solvents for processing	355 gal	Fuel blending
Debris (not specified)	-	Treated
Waste chromic acid solution	2,541 gal	Treated
Waste cyanide poison B solid	1 drum	Treated
Spent cyanide bath solution	6 drums	Treated
Potassium chromate waste	1 drum	Treated
Chromic acid mixture	1 drum	Treated

AEROQUIP/REPUBLIC HOSE, YOUNGSTOWN, OHIO (REGION V)

Site History

The site was a 9-acre abandoned industrial facility containing between 10 and 15 buildings. Many of these buildings were in a state of disrepair due to lack of maintenance, water leaks through broken windows, holes in roofs, and looting (scrap metal, electrical equipment).

The following materials were left on site:

- 300 drums of high-level PCB solids
- 200 drums of low-level PCB solids
- 10 drums of decontaminated fluids
- 8 drums of dielectric PCB oils
- 13 medium-sized transformer casings
- 1 large transformer
- 1 small capacitor

Contaminants

The principal contaminants found on site were PCB oils.

Site Description

Transformers on the roof of a five-story building had been tipped over and released PCB oils onto the asphalt roof and the gutters and into the building interior. Substantial soil contamination and possible contamination of sediments were also noted in a nearby creek. The following work was ordered to be performed at the site:

- Sampling of suspected areas of contamination
- Removal and disposal of all PCB-contaminated materials
- Decontamination of PCB-contaminated articles and disposal of all cleanup material

Equipment

Transformer casings, a transformer, and a capacitor were located inside a five-story building. Transformer casings on the second and third floors were manually carried to access windows and lowered to ground level by use of a backhoe and support chain. The larger transformer was removed from the boiler house by using cutting torches and a 6-ton towmotor. The cases and transformer were staged in a plastic-lined diked area inside an adjacent building.

Cleanup of the second-story roof of the manufacturing building was accomplished by using a towmotor to lift a dumpster up to the roof and loading it with the contaminated material (tarpaper, cement, gutters, window frames) to minimize contamination of other parts of the building.

Oil-soaked ash found in the sub-basement of the boiler house was removed by chopping it into small chunks (with picks and shovels) and pushing it into the hose of a vacuum unit (Supersucker). All PCB-contaminated material was staged in two piles (high or low PCB contamination) and covered with Visqueen.

Cleanup of the sewer system involved using two trash pumps to reroute the water. A Supersucker vacuum tanker was used to remove bottom sediments and surrounding soils. A backhoe was used to excavate and remove contaminated soil from the sewer outlet to the creek and to spread riprap to prevent soil erosion. Table 22 presents a detailed listing of the equipment used.

TABLE 22. EQUIPMENT USED AT THE AEROQUIP/REPUBLIC HOUSE SITE

Hydraulic drum grappler	Dump truck (5- to 10-yd^3)
Jack hammer	Trash pump (2-in.)
Generator (5-kW)	Hand tools
Backhoe (CAT 215)	Holding tank (300-gal)
Lowboy (20- to 40-ton)	Vacuum Tanker (1800 gal)
Portable holding tank	Forklift (2-ton)
Pressure pump (1.5-in.)	Forklift (6-ton)
Crane (20-ton)	Cutting torch
Circular saw	Air chipper
Backhoe (Case 580)	HP/HW washer (1200-psi)

Problems Encountered

Despite major removal efforts, few problems were encountered. Routine logistical problems involved in moving heavy equipment within a limited space were common. Because of the weight of the liquid-soaked material in the boiler house, the Supersucker hose became clogged and had to be replaced.

Final Disposition

Table 23 presents a breakdown of the amounts and disposition of wastes found on the Aeroquip/Republic Hose site.

TABLE 23. SUMMARY OF OFF-SITE DISPOSAL OF WASTE FROM THE AEROQUIP/REPUBLIC HOSE SITE

Waste type	Amount	Disposition
PCB-contaminated material	258 tons	Landfilled
PCB transformers	625 ft^3	Treated/landfilled
PCB transformers	425 gal	Treated/landfilled
PCB oil	12 gal	Incinerated
PCB capacitor	140 lb	Treated/incinerated
PCB-contaminated debris	7 drums	Landfilled

G&H LANDFILL, UTICA, MICHIGAN (REGION V)

Site History

This site is a 17-acre closed municipal and industrial landfill, which at one time received waste solvents, oils, grease, and automotive waste at volumes of up to 60,000 gallons/month. Liquid industrial waste was stored on site in lagoons. In 1982, a removal action was undertaken to prevent leaking of PCB-contaminated oils. In 1986, construction began on an interceptor/collector trench and a groundwater barrier made of sheet piling.

Contaminants

The principal contaminants found on the site were as follows:

PCBs	Xylenes (1,2; 1,3; 1,4-dimethylbenzene)
Ethylbenzene	Zinc
Toluene	Copper
Phenols	Cyanide
Lead	Mercury

Site Description

To prevent public access and further migration of PCB-contaminated oils, the following actions were undertaken:

- Recovery and storage of PCB oils
- Groundwater barrier construction
- Increased site security and restriction of access

Equipment

To collect/contain isolated seepages, a series of "pipe dams" were constructed to form a long shallow collector/interceptor trench. A trackhoe and a perforated bucket were used to excavate the trench, which allowed the liquids to drain while retaining the solids and debris. The liquids remaining in the trench were pushed to the outlet by using a trash pump and hose nozzle attached to the trackhoe bucket. Water was sprayed from the nozzle to push the viscous liquid and to wash the banks of oil. A vacuum truck was used to pick up liquids and PCB oils.

The groundwater barrier was positioned downgradient from the interceptor trench. The barrier was constructed of 8-ft-long sheets of steel piling driven 7 ft into the ground by use of a crane-mounted, vibrating, pile driver. A detailed equipment list is provided in Table 24.

TABLE 24. EQUIPMENT USED AT THE G&H LANDFILL, SITE

Front-end loader (CAT 955)	Dump truck (10- to 20-yd^3)
D-3 bulldozer (wide-track)	Chain saw
Backhoe (Case 580)	Trash pump (3-in.)
Backhoe (Cat 225)	CAT 215 bucket
Backhoe (CAT 215)	Submersible pump (1-in.)
Crane (20-ton)	Circular saw
Pile driver (V-5)	Vacuum truck (3500-gal)
Tractor, diesel w/40-ton lowboy	Pressure washer (3000-psi)
Dump truck (5- to 10-yd^3)	

Problems Encountered

The viscous nature of the oil mixed with the debris in the interceptor trench impeded the vacuuming of the oil. Attaching the hose nozzle of the trash pump to the trackhoe bucket made it easier to move the oils in the trench. Because of the presence of a wetland area containing heavy vegetation, extensive site preparation was necessary before the removal action could begin. The site also had to be kept as dry as possible for heavy equipment maneuverability.

Final Disposition

Contaminated liquids were stored on site. A pole barn was constructed to hold three 5800-gallon plastic storage tanks for temporary storage of these liquid wastes until their disposal offsite.

PBM ENTERPRISES, ROMULUS, MICHIGAN (REGION V)

Site History

This site, an abandoned silver recovery facility, contained X-ray film that had been left on site. Approximately 400 yd^3 of used film and chips were found in steel drums stored in dilapidated 40-foot semitrailers. Also present on the site was miscellaneous debris, including old truck tires and wooden pallets.

Contaminants

The principal contaminant found on site was cyanide.

Site Description

For the mitigation of potential threats posed by the hazardous waste stored at this site, a two-phase immediate removal action was begun. The first phase involved providing site security (to prevent direct exposure) and minimizing leachate and subsequent runoff. The second phase involved treatment with sodium hypochlorite (to oxidize the sodium cyanide) and disposal of the cyanide-contaminated chips.

Equipment

A pole barn was constructed to hold the two reaction vessels, pumps, valves, and a catwalk for observation and sampling. The reaction vessels were formed from rolloff boxes lined with PVC. Aeration and drain lines were built into the bottom of the vessel. Agitation of the material (chips and film) in the reaction vessel was accomplished by aeration from a large air compressor. A concrete drying pad was built, and a sump was dug to collect runoff/spillage.

Approximately 100 yd^3 of soil was excavated by use of a backhoe. A front-end loader was used to load the reaction vessels with contaminated chips. The 40-ft semitrailers were stripped of wood, and the wood was passed through the treatment process along with the steel drums. The semitrailers were vacuumed clean and returned to their owners. A backhoe was used to dredge the sediments contained in a nearby ditch, and a grader was used to spread them out on the site for treatment. A detailed equipment list is presented in Table 25.

Problems Encountered

Despite the large volume of liquids to be handled, materials-handling problems were relatively few. Although the PVC-lined rolloff boxes were adequate for the treatment process, leakage and spillage eventually occurred. Future treatment of this kind could lead to increased vendor interest in supplying secure leak-proof vessels.

Final Disposition

Table 26 presents a breakdown of the amounts and disposition of wastes found at the PBM site.

TABLE 25. EQUIPMENT USED AT THE PBM ENTERPRISES, SITE

Front-end loader (955 CAT)	Wet Vac
Generator (5-kW)	Lowboy (20- to 40-ton)
Front-end loader/crawler (CAT 936)	6000-gal tank
Backhoe (Case 580 or equivalent)	3000-gal tank
Tanker truck (5000-gallon)	Acid pump
Flatbed truck	Tanker truck (6000-gal, rubber)
Gradall	Dump truck
Compressor (175 cfm)	Chemical pump
Compressor (750 cfm)	Rolloff box
Pipe wrench (48-in.)	560 Steel tank

TABLE 26. SUMMARY OF OFF-SITE DISPOSAL OF WASTES FROM THE PBM ENTERPRISES, SITE

Waste type	Amount	Disposition
Treated film chips, debris, sand, and wood	1,280 yd^3	Landfilled
Contaminated runoff	5,564 gal	Treated
Alkaline corrosive liquid	26,572 gal	Treated
Alkaline waste	12,976 gal	Treated
Alkaline solid	4,240 gal	Treated
Alkaline liquid	147,483 gal	Treated
Alkaline wastewater	3,900 gal	Treated
Alkaline liquid treated	16,942 gal	Treated
Solvent-containing solids	82 yd^3	Landfilled

MIDCO II, GARY, INDIANA (REGION V)

Site History

The site is a 7-acre tract in an industrial area of Gary, Indiana, located in Lake County in the northwest part of the State. Until 1977, the site was used as a solvent, acid, base, and industrial waste recycling/disposal facility. In August 1977, a major fire completely burned the facility, and it was abandoned at that time.

Materials left on the site included the following:

°   Two large piles of burned-out drums (approximately 65,000) and approximately 2500 drums that were still intact

°   Twelve 10,000-gallon above-ground tanks and one buried tank; nine of the tanks still contained hazardous materials

○ Two areas used for dumping paints, sludges, plating waste, and PCBs (a sludge pit 120 ft x 30 ft x 10 ft, and a filter bed 150 ft x 40 ft x 10 ft)

○ Contaminated soil.

Contaminants

Contaminants found on site were as follows:

1,1,1-Trichloroethane	Methylene chloride	PCBs
Phthalates	Naphthalene	Cadmium
Phenols	Toluene	Lead
Ethylbenzene	Arsenic	Cyanide

Site Description

Many of the drums found at the site contained hazardous and/or flammable substances and were either leaking or in a deteriorated condition. The sludge pit and filter bed (which contained paints, sludges, plating waste, and PCBs) were unlined and therefore posed a threat to groundwater. The tanks contained flammable liquid waste, including solvents, paints, and paint sludge. The underground tank was believed to be leaking into the groundwater. Because of these perceived threats, an immediate removal action was recommended. The remedial action performed at this site took place during the period 1985 to 1988. The following work was ordered performed at the site:

○ Establishment of staging and loading areas
○ Conducting drum removal
○ Cleaning, removal, and disposal of on-site above- and below-ground tanks
○ Removal and disposal of onsite sludge
○ Installation of a clay cap
○ Air monitoring

Equipment

This site presented a major materials-handling effort involving several problems, including staging, separation, and disposal of numerous drums; excavation of underground tanks; decontamination of large above-ground tanks; excavation of the sludge pit; and extensive site preparation. A detailed equipment list is provided in Table 27.

TABLE 27. EQUIPMENT USED AT THE MIDCO II, SITE

Backhoe (Case 225)	Holding pool (300-gal)
Backhoe (Case 580)	Front-end loader (CAT 953)
Drum grappler (hydraulic)	High-pressure water laser
Stake-bed truck (2-ton)	Drum shredder
Diaphragm pump (3-in.)	Drum punch
Trash pump (2-in.)	Vacuum unit (1500-gal)
Generator (5-kW)	Cutting torch
Hand tools	Submersible pump
Lowboy (20-ton)	Holding pool (24-ft-diameter)
Diesel tractor	

Tracked backhoes outfitted with drum grapplers were used for the initial drum removal. The drums were sorted into liquid and solid waste piles. The drums containing solids were picked up with the grappler and fed into a drum shredder. A backhoe was then used to move the shredded material to an area where it was sampled before a tracked loader placed it onto a truck for off-site disposal. Liquid drummed waste was emptied into a vacuum truck and sent off-site for treatment.

A cutting torch was used to dismantle the aboveground tanks, and a backhoe was used to excavate the underground "clarifier tank." A large amount of scrap metal on the site had to be cut into smaller pieces and decontaminated for off-site disposal.

A backhoe was used to excavate the two sludge pits, and the material was moved to the treatment area with a rubber-tired loader. The sludge was solidified with lime.

Problems Encountered

During the 5-month effort, the ambient temperature ranged from -15 to 36° F. Because of these low temperatures, water (high-pressure laser) was not used to perform the decontamination. To avoid having to clean the heavy equipment, "clean roads" were built on site on which to run "clean equipment." Equipment used in the contaminated area stayed on-site until the effort was completed. An additional problem presented by the cold weather was the frequent shattering of air-line fittings. Heavy equipment was provided with extra lubrication to prevent breakdowns, and during the coldest weather, part of the crew arrived on-site half an hour early to warm up the equipment before actual work operations began.

Final Disposition

Table 28 presents a breakdown of the amounts and disposition of wastes found on the Midco II site.

TABLE 28. SUMMARY OF OFFSITE DISPOSAL OF WASTES FROM THE MIDCO II SITE

Waste type	Amount	Disposition
Cyanide soil/solids	3137 tons	Landfilled
PCB soil	7035 tons	Landfilled
Non-PCB liquids	60 tons	Incinerated
Cyanide mixture	6 drums	Treated
Tank-cleaning waste	5000 gal	Treated

MOTCO SITE, LA MARQUE, TEXAS (REGION VI)

Site History

This 11-acre site is an abandoned tar reclamation facility consisting of seven waste disposal pits and nine storage tanks containing 4 million gallons of liquids and styrene tar. The site is adjacent to a marsh that is used extensively for shrimping, crabbing, and fishing. Previous removal actions have involved the following:

o   Dike repairs, redrumming, removal and disposal of hazardous drums - 1981.

o   Carbon treatment of 2 million gallons of lagoon contents, and increased freeboard - 1981.

o   Lowering of water level in pits - 1983 and 1985.

Contaminants

Vinyl chloride   Acids
Styrene tars   Lead
Chlorinated hydrocarbons   Mercury

Site Description

Heavy rains in 1986 resulted in elevated water levels in the disposal pits. The dike was within 4 inches of overtopping and contaminating a nearby marsh. The following work was ordered to be performed:

o   Staging of pumping equipment and caustic material

o   Pumping water from Pit 1; injecting caustic into the line and discharging into a settling pond

o   Pumping treated water from the settling pond into roadside ditch

Equipment

Six-inch diesel pumps were used to pump out the pits. Table 29 shows equipment used on-site.

TABLE 29. EQUIPMENT USED AT THE MOTCO SITE

Diesel pump (6-in.)	Polyethylene tank
Air compressor	Hand tools
Generator	

Problems Encountered

Cold weather caused the caustic soda to become very viscous. A new valve was installed on the treatment manifold to improve flow.

Final Disposition

Approximately 1.5 feet of water was pumped out of the pit.

CLEVE REBER SITE, SORRENTO, LOUISIANA (REGION VI)

Site History

This 25-acre site is a former landfill that accepted both municipal and industrial waste. Numerous drums containing chemical waste were buried on the site. (Spillage of volatile chemical wastes occurred during the handling and disposal of the drums. The site contained four surface ponds, five major surface piles of metal drums, and a buried waste pit.) The adjoining land is a mix of swamps/dense vegetation and sparsely populated residential areas.

Contaminants

The contaminants of primary concern include the following:

4,4'-DDT	Di-n-butyl phthalate
4,4'-DDE	Hexachlorobutadiene
4,4'-DDD	Arsenic
Hexachlorobenzene	Aluminum
Mercury	Lead
Chlordane	Toluene
Chlorobenzene	Atrazine

Site Description

The majority of the industrial waste was disposed of in a pit that varied from 6 to 20 feet in depth and held approximately 220,000 yd^3 of waste. Leachate from the pit has migrated from the pit and mixed with the municipal waste. Also, ponds on the site contain about 22 million gallons of surface water. An estimated 6400 drums are buried on site, and refuse debris is scattered over the surface. Finally, the site has areas of stained soil and zones where black tarry substances leak to the surface through cracks in the soil.

The removal action ordered for this site involved removing the sources of surface contamination.

Equipment

The primary pieces of equipment used for the excavation of the drums were bulldozers and front-end loaders. Bulldozers were also used to clear vegetation and to build a dirt and sand access road. Drum crushing at the site was accomplished with bulldozers. Table 30 is a partial listing of the equipment used on the site.

TABLE 30. EQUIPMENT USED AT THE CLEVE REBER SITE

Bulldozer *	Backhoe
Dump trucks	Hand tools
Wheeled front-end loaders *	Rolloff box
PVC liners for trucks	

Problems Encountered

Muddy conditions from excess rain resulted in the bulldozer being used to push trucks onto and off of the site. The front-end loader also needed to be pushed out of the mud on occasion. An additional problem occurred because the drums were buried very close to the surface or, in one instance, above ground level with soil mounded over them. During the operation of the heavy equipment, care had to be taken to avoid disturbing these drums.

Final Disposition

During the removal action, 70 truckloads of waste and 1100 crushed drums were removed from the site, and 713 yd^3 of clay was brought in for capping of contaminated areas.

QUAIL RUN, GRAY SUMMIT, MISSOURI (REGION VII)

Site History

Quail Run Mobile Manor is a mobile home park in which the road through the park had been sprayed for dust control with 2,3,7,8-tetrachlorodibenzo-p-dioxin (TCDD)-contaminated waste oil. The affected area included:

- 28 mobile homes
- 1 house

- The roof of an underground home
- Saturation under an equipment shed
- 1400 feet of highway shoulder
- 5 acres of park property

Contaminants

The principal contaminant found on site was TCDD in concentrations as high as 2200 µg/kg.

Site Description

Twenty-one families were temporarily relocated during the site cleanup activities. The following work was performed at the site:

- Decontamination of on-site debris
- Excavation of contaminated soil
- Decontamination and restoration of 28 mobile homes
- Decontamination and restoration of a rental house

Equipment

Decontamination of the mobile homes and brick house included a series of vacuum, high-pressure wash, and wipe-down cycles. Porous fabric within the mobile homes was stripped out by hand and disposed of as contaminated material.

Soil excavation was accomplished with a backhoe. The backhoe was then used to load the soil into woven polypropylene bags with a polyethylene liner (2-yd^3 capacity). The bags were stored on site in an enclosed steel-sided storage building with an impermeable asphaltic concrete floor.

Bulldozers, backhoes, and pressure washers were used to decontaminate large chunks of concrete from patios and concrete pads. The concrete was suspended from a crane by a wire noose while being pressure washed with a detergent solution. Table 31 lists the equipment used.

TABLE 31. EQUIPMENT USED AT THE QUAIL RUN SITE

Bulldozers*	Highlift (CAT 936)
Tracked loader (CAT 943)	Generator (5-kW)
Wheeled loader (CAT 910)	Cranes (18- to 32-ton)
Hand tools	Tracked excavator (CAT 215)
High-pressure washer	Gradall
Backhoe (Case 580)	Dump trucks
Aluminum hopper	Air compressor
Garden hose	Fire hose (2-in.)
Bag platform	Hopper

The following heavy equipment (with function description) was regularly used during the excavation phase:

CAT 215	-	A large-capacity trackhoe, equipped with a 48-in. straight-edged bucket to obtain a smooth cut. One CAT 215 was equipped with an articulated bucket suitable for cutting at different slope angles.
Case 580	-	A smaller capacity, rubber-tired backhoe, equipped with a straight-edged bucket.
CAT 910	-	An articulated highloader that was primarily used for transport of supplies and materials to the work crews, as well as for movement of concrete.
Cranes	-	Two to three cranes of varying capacities (18- to 32-ton) were used for removing bags from the hopper and for loading and unloading bags into dump trucks. Cranes were also used for suspending concrete for decontamination.
Hopper	-	A custom-designed and fabricated aluminum funnel on a stand for holding the bulk bags and channeling dirt into the bags from the backhoe bucket.
Trucks	-	Dump trucks were used during earlier excavation phases for transport of bags within the contaminated area of the park and, primarily, within the clean areas. Dump trucks were also used for transport of bags of soil.
CAT 936	-	A high-capacity, extending highlift equipped with custom-designed and manufactured forks made from heavy-gauge, 6-in.-diameter steel pipe. This highlift was used for removing full bulk bags from the trucks and stacking them within the storage structures.
Bag platform	-	This stand was custom designed for holding a full bag while the forks of the CAT 936 were placed under the bag for lifting it to the top row of the stack.

Problems Encountered

Continued rainfall and muddy conditions caused the project to be shut down for several months. The mud caused excavation problems as well as problems with loading soil/mud into the bags. Additional problems occurred during excessively dry weather when it became necessary to apply water to the excavation area as a dust-control measure. Rubber-tired vehicles were primarily used instead of tracked ones to minimize the mixing of surface-contaminated soils.

Final Disposition

The following material was disposed of or decontaminated at the site:

- Six hundred cubic yards of concrete
- Approximately 10,121 two-cubic-yard bags of soil
- Approximately 2101 two-cubic-yard bags of contaminated debris
- Eight 20-yd^3 dumpsters of decontaminated debris

SOLID STATE CIRCUITS, REPUBLIC, MISSOURI (REGION VII)

Site History

This 1/2-acre site is a former cold-storage warehouse (refrigeration plant) and circuit board manufacturing facility. During the manufacturing of circuit boards, approximately 42 drums of trichloroethylene (TCE) were used over a period of 5 years, resulting in the accumulation of 2310 gallons of liquid waste that was stored in a basement sump. In February 1979, the building burned to the ground. Debris from the fire was bulldozed into the basement of the building. A cap was not placed over the debris, and water was able to percolate freely over the debris and fill material in the basement. In 1984, the Missouri Department of Natural Resources (MDNR) and the manufacturer conducted soil and liquid sampling around the basement and excavated soil and debris from the basement area that has been stockpiled on site.

Contaminants

The principal contaminant found on site was TCE in concentrations as high as 460,000 ppb (soil).

Site Description

The city of Republic's water supply was found to be contaminated with concentrations of TCE between 23 and 140 ppb. Debris and soil from the 1984 excavation were still stockpiled on the site. The following work was ordered to be performed at the site:

- Loading and transportation of 1400 yd^3 of soil that had been stockpiled during the SSC/MDNR removal action of 1984

- Removal of an additional (up to) 880 yd^3 of contaminated soil

- Development and construction of six monitoring wells at the site. Four wells were to be shallow (18 to 20 feet deep), one was to be approximately 400 feet deep, and another was to be 600 feet deep

- Removal of contaminanted shallow groundwater. Disposal of the contaminated water at the Springfield, Missouri, municipal wastewater treatment plant

Equipment

Excavation of the concrete slab in the basement required the use of a pavement breaker (hoe ram) attachment for the backhoe because of the thickness of the slab (14 to 18 inches). Table 32 lists the equipment used.

TABLE 32. EQUIPMENT USED AT THE SOLID STATE CIRCUITS SITE

Equipment	Utilization
7500-gal galvanized tanker	Hauled contaminated water
6000-gal galvanized tanker	Hauled contaminated water
4.5-yd^3 front-end crawler (loader)	Excavation and loading of contaminated soil
Backhoe (Case-580)	Excavation
Emergency-response van	Command post and equipment storage
Backhoe (CAT-255)	Excavation
Pickup truck	Hauled supplies
Passenger sedan	Transported workers to and from site
Weight scales	Weighed trucks
Tractor trailer	Mobilization and demobilization
Pumps	Pumped wastewater
Compressor	Operation of air hammer
Air hammer	Cracked concrete slab
Hoe-ram attachment	Cracked concrete slab and footings
Stake-bed truck	Equipment transportation
Cutting torch	Cutting of basement pipes and elevator shaft

Problems Encountered

The only major materials-handling-related problems encountered were project delays resulting from the unexpected problems of having to obtain a pavement-breaking attachment for the backhoe (because of the thickness of the concrete slab) and the tank trucks for pumping and hauling the contaminated water accumulated during the excavation.

Excavation stopped during the project because of a problem with the offsite facility chosen for soil disposal. This resulted in site activities being suspended for more than 3 months. When excavation resumed several thousand gallons of water had filled the excavation area, and tanker trucks were brought in to pump out the contaminated water.

Final Disposition

Table 33 presents a breakdown of the amounts and disposition of wastes found at the Solid State Circuit site.

TABLE 33. SUMMARY OF OFFSITE DISPOSAL OF WASTE FROM THE SOLID STATES CIRCUIT SITE

Waste type	Amount	Disposition
TCE-contaminated water	108,000 gal	Wastewater treatment plant
TCE-contaminated soil	1,990 tons	Landfilled

B&C METALS, DENVER, COLORADO (REGION VIII)

Site History

A 33,000-ft^2 building housed a former radium-processing operation. The elevated radon levels found in the building were the result of infiltration of radon gas from nearby soil (2400 yd^2) contaminated with radium (Ra-226). The building was still in use at the time of the removal action.

Contaminant

The contamination on the site resulted from the decay of radium to radon gas, which is a naturally occurring radioactive gas.

Site Description

The building includes a first-floor shop and office area, a basement storage area, and a basement appliance-refinishing operation. The work ordered to be performed was the installation of a plenum wall-stack vent system to reduce radon levels.

Equipment

Table 33 presents a summary of the equipment used on the site. The work performed dealt primarily with the installation of the vent and sealing the building against the entry of any radon gas. Table 34 lists the equipment used.

TABLE 34. EQUIPMENT USED AT THE B&C METALS SITE

Pickup truck	Submersible pump
Wheelbarrow	Generator
Cutting torch	Tractor
Hand tools	

Problems Encountered

No materials-handling problems were reported.

Final Disposition

No waste was removed from the site, but radon levels in the building were reduced to levels deemed acceptable to the EPA.

BURLINGTON NORTHERN RAILROAD, SOMERS, MONTANA (REGION VIII)

Site History

The site was formerly a facility that treated railroad ties with wood preservatives. The preservatives included a zinc chloride solution and a creosote/diesel oil mixture. The northern end of the site was used as a storage area for treated and untreated wood and process residues. From 1901 to 1971 the plant discharged zinc and creosote-contaminated liquid and solid wastes to a disposal lagoon, which occasionally overflowed into a drainage ditch that emptied into a nearby lake.

Contaminants

The main contaminants of concern include:

- Naphthalene
- Phenanthrene
- Acenaphthene
- Fluorene
- Benzo(a)anthracene
- Benzo(a)pyrene
- Pyrene
- Zinc
- Nickel
- Copper
- Arsenic

Site Description

A small pond, created from overflow of the disposal lagoon into a topographically low area, (approximately 200,000 gallons) just north of the Flathead Lake shoreline contains sludges that are heavily contaminated with creosote-related compounds to a depth of 1 to 2 feet. The following work was ordered to be performed at the site:

- Draining the standing water in the swamp pond and depositing the water into either a 150,000-gallon storage tank on the BN property or a lined (60-mil) RCRA overflow pit on the BN property

- Excavation and removal of the contaminated soils and sludges from the drained pond and depositing them in a lined (60-mil) RCRA sludge pit

- Backfilling the resulting cavity to a level that matches the existing ground surface

Equipment

Submersible pumps were used to pump the contaminated water into the RCRA pit and the storage tank. A working ramp for the heavy equipment was constructed to provide access to the sludge on the bottom of the pond. The ramp required 900 yd³ of fill. Table 35 lists the equipment used on the site.

TABLE 35. EQUIPMENT USED AT THE BURLINGTON NORTHERN SITE

Backhoe*	Submersible pump
Front-end loader (4.5-yd³ bucket)	Discharge hoses
Bulldozer (CAT D3)	Rolloff box
Polyethylene storage tank	Hand tools

Problems Encountered

The only major materials-handling problem encountered involved continued pumping of contaminated water from the pond area as a result of rains and recharge.

Final Disposition

Total contaminated water discharge to the overflow pit and the storage tanks was 127,000 gallons. Total sludge and soil removed to the RCRA pit was 3280 yd³.

McCOLL SUPERFUND SITE, FULLERTON, CALIFORNIA (REGION IX)

Site History

The McColl site was used for disposal of acidic refinery sludge, a byproduct of the production of high-octane aviation fuel in the early and mid-1940's. In the 1950's, fill material and drilling muds from nearby oil exploration activities were also deposited there. Later, the site was covered with soil in an attempt to make it suitable for development.

Contaminants

Three major waste types are present at the site: a green mud-like material, a black viscous tar-like material, and a black char-like asphaltic waste. The principal air emissions of concern are sulfur dioxide (SO_2) and volatile organic compounds. The following is a detailed list of the chemical constituents known to be present at the site:

Methylene chloride	Aluminum
Acetone	Barium
2-Butanone	Calcium
Benzene	Chromium
Toluene	Copper
Ethylbenzene	Cobalt
Xylene	Lead
Vanadium	Manganese
Zinc	

Site Description

The 20-acre site contains 12 waste pits or sumps. A trial excavation of part of one sump was performed over a 6-week period with approximately 100 cubic yards of waste being removed. This trial excavation and waste handling project was performed for a better definition of potential air-emission and materials-handling problems expected during full-scale excavation and treatment. The potential impact of air emissions on the local community was also investigated.

The work was performed within a temporary enclosure, and air was exhausted from the enclosure through a sodium-based wet scrubber and a granular activated-carbon-bed adsorber. Foam was used to suppress atmospheric releases from the raw waste excavation, storage, and processing. The air exhaust from the enclosure was monitored for total hydrocarbons (THC) and SO_2. These and other selected organic and reduced sulfur compounds were monitored along the site's perimeter and in the ambient air of the surrounding neighborhood.

Equipment

A 60 ft x 160 ft, rigid-frame, polyvinylchloride-covered enclosure was constructed to contain all air emissions released during the trial excavation. Ventilation air was continuously drawn through the enclosure at a rate of 1000 ft^3/min. An induced-draft fan operating outside the enclosure directed the ventilation air to the air pollution control system and exhaust stack.

Excavation of the 15-ft-long x 10-ft-wide x 25-ft-deep pit was accomplished with a track-mounted backhoe. A 20 ft x 20 ft working pad was constructed for the trackhoe operation. A trench shield constructed from 1/4-inch carbon steel was used to shore the walls of the excavation pit.

The excavated material was categorized by waste type and transported to individual waste piles within the enclosure by a front-end loader. The excavated char and mud wastes did not require processing. The tar waste was solidified with lime-based additives in a pug mill consisting of two shafts fitted with short paddles mounted on a cylinder. After mixing, this material was extruded and automatically cut into small pellets. Samples of the different wastes were collected in 55-gallon drums for future testing as feed material for a thermal destruction technology.

After processing was completed, all remaining materials were returned to the pit. Refilling and restoration of the pit were accomplished with a front-end loader. Table 36 lists of the equipment used at the McColl site.

TABLE 36. EQUIPMENT USED AT THE McCOLL SITE

Backhoe*	Fan
Soil crusher	Vibratory screen
Forklift	Rolloff box
Pickup truck	Pug mill
Skid-steer loader	Pressure washer
	Generator (40-kW)

Problems Encountered

Despite the application of a foam vapor suppressant, much higher than expected concentrations of SO_2 and THC were generated within the enclosure. The foam did not adhere well to the waste and tended to degrade faster than expected. The high concentrations required that excavation work be performed in Level A personal protective equipment.

Seepage of tar into the excavation pit occurred despite the use of a trench shield. This caused some excavation difficulties.

PACIFIC HIDE AND FUR, POCATELLO, IDAHO (REGION X)

Site History

The 10-acre site is a former metals reclamation operation. The facility overlies an aquifer that serves as a drinking water source for local residents. Material left on the site included the following:

- An unknown number of transformer parts
- Approximately 582 capacitors
- 650 drums
- Large pieces of debris (vehicle frames, washers, dryers, refrigerators, bins, tanks, combine chassis, cars)

Contaminant

The primary contaminant of concern were PCBs.

Site Description

The capacitors and transformer parts on the site were leaking PCB-contaminanted oil onto the ground. Many of the drums were in a deteriorated condition and leaking. A Federal search warrant was necessary to gain access to the site. During two separate removal actions (1983 and 1984) the following work was performed:

- Contaminated soil was excavated and disposed of.
- Capacitors were staged and shipped out for disposal.
- Drums were overpacked and shipped out for disposal.
- Large pieces of scrap were cleaned.

Equipment

A backhoe and front-end loader were used to excavate the soil. A crane was used to stage and load the capacitors into overpacks. A steel work pad, sump, and curtain structure were constructed to steam-clean and pressure-wash (with detergent) the large pieces of scrap found on the site. Table 37 shows the equipment used at the site.

TABLE 37. EQUIPMENT USED AT THE PACIFIC HIDE AND FUR SITE

Steel drum roller	High-pressure/hot-water washer
Backhoe*	Crane (15-ton)
Front-end loader*	Forklift
Bulldozer*	

Problems Encountered

No major materials-handling problems were encountered at this site.

Final Disposition

Table 38 presents a breakdown of the amounts and disposition of wastes found at the Pacific Hide and Fur site.

TABLE 38. SUMMARY OF OFF-SITE DISPOSAL OF WASTES FROM THE PACIFIC HIDE AND FUR SITE

Waste type	Amount	Disposition
Capacitors	582	Incinerated
Contaminated soil	30 yd^2	Landfilled
Drums	16	Landfilled
Debris	180 large pieces	Steam-cleaned and left on site

NORTHWEST TRANSFER SALVAGE YARD, EVERSON, WASHINGTON (REGION X)

Site History

This 1.2-acre site is a former transformer storage and recycling facility. The open unsecured site contained 200 transformer casings on the ground. The site also contained the following:

- ○ Oil-stained soil patches around the transformers
- ○ An unlined air curtain pit incinerator
- ○ A barn used for transformer disassembly

Contaminants

The primary contaminant of concern were PCBs.

Site Description

Sampling revealed the presence of 35 PCB-contaminanted transformers, soil contamination, and PCB contamination in the barn. The following work was ordered performed:

- Cleaning and rinsing of transformers contaminanted with PCB fluids
- Cleanup and decontamination of the barn
- Excavation and disposal of contaminated soil

Equipment

The following protocol and equipment were used to clean the transformers:

1. Transformers were staged above a large, galvanized metal stock water tank with the aid of a forklift.

2. The PCB-contaminated oil was drained into the tank and then pumped to the appropriate compartment of a bulk tank truck staged adjacent to the work area. The truck was divided into four compartments; two were designated for PCB fluids ranging from 0 to 45 ppm, and the other two were designated for liquids containing 45 to 500 ppm PCBs.

3. Transformers were rinsed three times, each time with a volume of diesel fuel equivalent to 10 percent of the total transformer volume.

4. Rinsings were drained and pumped to the tank truck after each flushing.

During the barn decontamination, a jack hammer was used to remove a heavily contaminated concrete berm. The concrete floor was pressure-washed and steam-cleaned with a high-pressure/hot-water washer. Wood on the inside of the building was sand-blasted.

Excavation of the soil was performed with bulldozers and backhoes. Six inches to one foot were removed from the area around the barn and incinerator. Soil around the septic system basin was removed to a depth of 20 feet. Table 39 lists the equipment used for the cleanup of this site.

TABLE 39. EQUIPMENT USED AT THE NORTHWEST TRANSFORMER SITE

Forklift	Air compressor
Steam/pressure washer	Jack hammer
Dumpster	4-wheel drum dolly
Stake-bed truck	Backhoe*
Trash pump (2-in.)	Bulldozer*
Generator (5 kW)	Clam shell crane
Electric barrel pump	

Problems Encountered

No major materials-handling problems were reported at this site.

Final Disposition

Table 40 presents a breakdown of the amounts and disposition of wastes from the Northwest site.

TABLE 40. SUMMARY OF DISPOSITION OF WASTE FROM THE NORTHWEST TRANSFORMER SITE

Waste type	Amount	Disposition
Soil	1400 yd^3	Landfill
Debris	Several piles	Landfill
High concentration PCB liquids	4660 gal	Incinerated
Low-concentration PCB liquids	1500 gal	Recycled

References

"ATSDR-EPA List of Second 100 Substances for Which Toxicological Profiles Will Be Developed [53 FR 41280, October 20, 1988]." Chemical Regulation Reporter, 1131, October 21, 1988.

Averett, D. E., B. D. Perry, and E. J. Torrey. 1989. Review of Removal, Containment, and Treatment Technologies for Remediation of Contaminated Sediment in the Great Lakes. Prepared for the U.S. Environmental Protection Agency by the Department of the Army, Corps of Engineers Waterways Experiment Station.

Bendersky, D., et al. 1980. Processing Equipment for Resource Recovery Systems, Volume 1, State of the Art. EPA-600/2-80-007A.

Bonner, T. A., et al. 1981. Engineering Handbook for Hazardous Waste Incineration. Prepared by Monsanto Research Corp. for U.S. Environmental Protection Agency. EPA SW-889.

Brunner, D. R., and D. J. Keller. 1972. Sanitary Landfill Design and Operation. U.S. Environmental Protection Agency, SW-65.

Church, H. K. 1981. Excavation Handbook. McGraw Hill Book Co., New York.

Cullinane, M. J., L. W. Jones, and P. G. Malone. 1986. Handbook for Stabilization/Solidification of Hazardous Wastes. Army Engineers Waterways Experiment Station. EPA/540/2-86/001.

Doerr, T. B., M. C. Landin, and C. O. Matrin. 1986. U.S. Army Corps of Engineers Wildlife Resources Management Manual, Sec. 5.7.1, Mechanical Site Preparation Techniques. Army Engineer Waterways Experiment Station, Report No. WES/TR/EL-86-17.

EG&G Idaho, Inc. 1984. Low-Level Radioactive Waste Technology - National Low-Level Radioactive Program. Prepared for the U.S. Department of Energy.

Esposito, M. P., et al. 1985. Guide for Decontaminating Buildings, Structures, and Equipment at Superfund Sites. Prepared by PEI Associates, Inc., and Battelle Laboratories for the U.S. Environmental Protection Agency. EPA/600/2-85/028.

"First Priority List of Hazardous Substances Subject of Toxicological Profiles by EPA and ATSDR [52 FR 12866, April 17, 1987]." Chemical Regulation Reporter, 125, April 24, 1987.

Hall, J. E. 1988. Application of Sewage Sludge to Agricultural Land. A Directory of Equipment. Medmenham Laboratory, United Kingdom.

John Deer and Co. 1989. Brochure for Remote Operated Heavy Equipment.

JRB Associates, Inc. 1982. Handbook for Remedial Action at Waste Disposal Sites. Prepared for the U.S. Environmental Protection Agency under Contract No. 6-82-006.

Kulwiec, R. A. 1985. Materials Handling Handbook. John Wiley & Sons, New York.

Malloy, C. W. 1984. Transportation of Low-Level Radioactive Waste. Presented by Westinghouse Hittman Nuclear, Inc., at the 11th Annual Energy Conference and Exhibition, February 22, 1984, Knoxville, Tennessee.

National Technology Seminar Proceedings. 1988. U.S. Environmental Protection Agency. Panama City, Florida. November 14-18, 1988.

Noble G. 1976. Sanitary Landfill Design Handbook. Technomic Publishing C., Inc., Westport, Connecticut.

Pasha Publications, Inc. 1989. 1989 Guide to Superfund Sites, Arlington, VA.

PEI Associates, Inc. 1989. Summary Report of Asbestos-Abatement Project Conducted in July 1988 at Western Hills High School. Prepared for Cincinnati Public School District.

Perry, R. H. 1984. Perry's Chemical Engineer's Handbook. McGraw-Hill Book Co., New York.

Pit and Quarry Handbook and Buyers Guide. 1975/76. Pit and Quarry Publications, Chicago.

Powerscreen of America, Inc. 1988. Materials Handling. Paper presented at U.S. Environmental Protection Agency National Technology Seminar, Panama City, Florida, November 14-18, 1988.

Raymond, G. L. 1984. Techniques to Reduce the Sediment Resuspension Caused by Dredging. Miscellaneous Paper HL-84-3. U.S. Army Corps of Engineers Waterways Experiment Station, Vicksburg, MS.

Robinson, W. D. 1986. The Solid Waste Handbook. Wiley Interscience, New York.

R. S. Means Company, Inc. 1988. Building Construction Cost Data 1989. R. S. Means Publishing, Kingston, Massachusetts.

Runyon, K.G. 1985. Improvement of Magnetically Separated Ferrous Concentrate by Shredding: A Performance Test. Prepared by the National Center for Resource Recovery, Inc., for the U.S. Environmental Protection Agency. EPA-600/2-81-103.

Savage, G. M., and G. R. Shiflett. 1980. Processing Equipment for Resource Recovery Systems. Vol. 3. Field Test Evaluation of Shredders. Prepared by California Recovery Systems, Inc. for the U.S. Environmental Protection Agency. EPA-600/2-80-007c.

Tamm, A. H., J. O. Cowles, and W. F. Beers. 1988. Investigation of Feedstock Preparation and Handling for Mobile On-Site Treatment Technologies. Prepared by Roy F. Weston, Inc., for U.S. Environmental Protection Agency under Contract No. 68-03-3450.

Tchobanoglous, G., H. Theisen, and R. Eliassen. 1977. Solid Wastes: Engineering Principles and Management Issues. McGraw-Hill Book Co., New York.

Todd, R., et al. 1988. Dust and Vapor Suppression Technologies for Excavating Contaminated Soils, Sludges, and Sediments. Prepared by Roy F. Weston, Inc., for U.S. Environmental Protection Agency under Contract No. 68-03-3450.

U.S. Army Corps of Engineers. 1986. Guidelines for Selecting Control and Treatment Options for Contaminated Dredged Material. Prepared by the U.S. Army Corps of Engineers Hydraulics Laboratory, Waterways Experiment Station, Vicksburg, MS. for the U.S. Environmental Protection Agency, Region 10.

U.S. Environmental Protection Agency. 1985a. Remedial Action at Waste Disposal Sites, Handbook. EPA/625/6-85-086a.

U.S. Environmental Protection Agency. 1985b. Guidance for Controlling Asbestos-Containing Materials in Buildings. EPA/560/5-85-024.

Wagner, K., et al. 1986. Drum Handling Practices at Hazardous Waste Sites. Prepared by JRB Associates, Inc., for U.S. Environmental Protection Agency. EPA/600/2-86/013.

Ware, S. A., and G. S. Jackson. 1978. Liners for Sanitary Landfills and Chemical and Hazardous Waste Disposal Sites. Prepared by Ebon Research Systems for U.S. Environmental Protection Agency. EPA-60019-78-005.

Bibliography

Beck, E. C. 1984. Specifiers Guide to Pump Selection. Pollution Engineering, 16(7):45-52.

Bendersky, D., and B. Simister. 1979. Research and Evaluation of Solid Waste Processing Equipment. In: Municipal Solid Waste: Resource Recovery, Report No. EPA-600/9-79-0-231b.

Berkiwitz, J. B., et al. 1976. Physical, Chemical, and Biological Treatment Techniques for Industrial Wastes, Volumes 1 and 2. Prepared by Arthur D. Little, Inc., for U.S. Environmental Protection Agency. EPA/SW-148C.

Camp Dresser and McKee Federal Programs Corp. Undated. Most Frequently Identified Compounds/Metals at Hazardous Waste Sites. Prepared by for U.S. Environmental Protection Agency under EPA Contract No. 68-01-6939.

Conway, R. A., and R. D. Ross. 1980. Handbook of Industrial Waste Disposal. Van Nostrand Reinhold Co., New York.

Dunning, P. B. 1981. Crushing Review. National Sand and Gravel Association Plant Operator's Forum.

Earth Technology Corp. 1988. Lessons Learned at Hazardous Waste Sites Solidification/Stabilization Processes. Prepared for the U.S. Environmental Protection Agency under Contract No. 68-03-3413.

Eastman, R. M. 1987. Materials Handling. Marcel Dekker Inc., New York.

Furman, C., et al. 1988. Case Studies Addendum: 1-8. Remedial Response at Hazardous Waste Sites. Prepared by Science Applications International Corp. for U.S. Environmental Protection Agency, 540/2-88/001.

Hatayama, H. D., et al. 1980. A Method for Determining the Compatibility of Hazardous Wastes. Prepared for the U.S. Environmental Protection Agency. EPA-600/2-80-76.

Hill, R. D. 1987. SITE Program-Superfund Innovative Technology Evaluation-- New Approaches to Cleaning Up Hazardous Waste Sites." Prepared for the U.S. Environmental Protection Agency. EPA/600/D-87/238.

Hill, R. D. 1987. Superfund. Prepared for the U.S. Environmental Protection Agency. EPA/600/D-87/293.

Kalu, A. O. 1987. Fundamentals of Industrial and Commercial Debris Management--An Economic Approach. Prepared for the U.S. Environmental Protection Agency.

Lundkuist, R. G. T. 1985. Handbook of Materials Handling. John Wiley & Sons, New York.

McCoy, D. E. 1985. Alternatives for On-Site Solidification/Stabilization. Hazardous Waste Consultant, 3(6):1-2,-1-7, November/December 1985.

Meade, J. P., and W. D. Ellis. 1985. Decontamination Techniques for Mobile Response Equipment Used at Waste Sites (State-of-the-Art Survey). Prepared by JRB Associates, Inc., for the U.S. Environmental Protection Agency. EPA/600/2-85/105.

Melvold, R. W., and L. T. McCarthy. 1984. Emergency Response Procedures for Control of Hazardous Substance Releases. Prepared by Rockwell International for the U.S. Environmental Protection Agency. EPA-600/D-84-023.

Nawrocki, M. A. 1976. "Removal and Separation of Spilled Hazardous Materials From Impoundment Bottoms." Prepared for Hittman Associates, Inc., for the U.S. Environmental Protection Agency, Report No. EPA/600/2-76/245. September 1976.

Nunn, N. 1989. Defining Hazwastes: Specialized Equipment. World Wastes, April 1989, pp. 106-109.

Olson, D. L. 1981. Screening Material. National Sand and Gravel Association Plant Operator's Forum.

PEI Associates, Inc., and Earth Technology Corp. 1989. Stabilization/Solidification of CERCLA and RCRA Wastes. Prepared for U.S. Environmental Protection Agency under Contract No. 68-03-3413.

Peterson, C., and D. Wilson. 1988. Shredding Wastes Before On-Site Remediation. Hazardous Waste Management, April 16-17, 1988.

Rosbury, K. D. 1985. Dust Control at Hazardous Waste Sites, Handbook. Prepared by PEI Associates, Inc., for U.S. Environmental Protection Agency. EPA/540/2-85/03.

Savage, G. M. 1979. "Evaluation and Performance of Hammermill Shredders Used in Refuse Processing." In: Municipal Solid Waste: Resource Recovery. EPA-600/9-79-023b.

Savage, G. M., et al. 1983. Significance of Size Reduction in Solid Waste Management. Vol. 3. Effects of Machine Parameters on Shredder Performance. Prepared by California Recovery Systems, Inc., for U.S. Environmental Protection Agency. EPA-600/2-83-006.

Scholz, R., and J. Milanowski. 1983. Mobile System for Extracting Spilled Hazardous Materials From Excavated Soils. Prepared by Rexnord Inc. for U.S. Environmental Protection Agency. EPA-600/2-83-100.

Institute of Scrap Recycling Industries, Inc. 1988. Scrap: America's Ready Resource.

Shultz, D. W., and D. Black. 1981. Municipal Solid Waste: Resource Recovery. In: Proceedings of the Annual Research Symopsium. Prepared by Southwest Research Institute for U.S. Environmental Protection Agency. EPA-600/9-81-003c.

Uterberg, W., et al. 1987. Reference Manual of Countermeasures for Hazardous Substance Releases. Prepared by Environmental Monitoring Services, Inc., for U.S. Environmental Protection Agency. EPA/600/2-87/069.

U.S. Environmental Protection Agency. 1984. Hazardous Waste Sites: Descriptions of Sites on Current National Priorites List. EPA/HW-8.5.

U.S. Environmental Protection Agency. 1984. Remedial Response at Hazardous Waste Sites: Summary Report. EPA-540/2-84-002a.

U.S. Environmental Protection Agency. 1985. Remedial Action at Waste Disposal Sites, Handbook. EPA/625/6-85/086.

U.S. Environmental Protection Agency. 1987. A Compendium of Technologies Used in the Treatment of Hazardous Waste. EPA/600/2-87/001.

U.S. Environmental Protection Agency. 1987. International Conference on New Frontiers for Hazardous Management. EPA/600/9-87/018F.

U.S. Environmental Protection Agency. 1987. Patterns of Soil Contamination and Composition on NPL Sites. Draft of an in-house report prepared by EO-EERU for EPA-Releases Control Branch.

U.S. Environmental Protection Agency. 1988. Technology Screening Guide for Treatment of CERCLA Soils and Sludges. EPA/540/2-88/004.

U.S. Environmental Protection Agency. 1988. The Superfund Innovative Technology Evaluation Program: Technology Profiles. EPA/546/5-88/003.

VERSAR. 1985. Capacity and Capability of Alternatives to Land Disposal for Superfund Wastes: Waste Type and Quantity Projections. EPA Contract No. 68-01-7053.

Walsh, J. 1987. Process Design Manual: Municipal Sludge Landfills. Prepared by SCS Engineers for U.S. Environmental Protection Agency. EPA/625/1-78/010.

Wetzel, R., and K. Wagner. 1983. "Drum Handling Practices at Abandoned Sites." In: Land Disposal of Hazardous Waste. EPA-600/9-83-018.

Wilder, I. 1982. Emergency Response Equipment to Clean Up Hazardous Chemical Releases at Spills and Uncontrolled Waste Sites. Prepared for the U.S. Environmental Protection Agency. EPA-600/D-82-348.

Wilson, D. G., ed. 1977. Handbook of Solid Waste Management. Van Nostrand Reinhold Co., New York.

Appendix A

Frequency of Occurrence of Contaminants at 1035 Superfund Sites

FREQUENCY OF OCCURRENCE OF CONTAMINANTS FOUND AT 1035 SUPERFUND SITES - 1989[a]

FREQUENCY OF OCCURRENCE

CONTAMINANTS	EPA REGIONS										TOTAL
NUMBER OF SITES	I	II	III	IV	V	VI	VII	VIII	IX	X	
	65	162	150	134	223	59	54	35	103	50	1035
TRICHLOROETHYLENE	23	39	44	15	51	7	8	3	46	10	246
LEAD	12	25	37	44	45	17	13	11	9	17	230
CHROMIUM	12	20	22	39	38	11	8	4	7	12	173
POLYCHLORINATED BIPHENYLS (PCBs)	18	30	15	23	40	12	2	1	6	9	156
HEAVY METALS		36	13	27	35	11	4	8	8	5	147
TETRACHLOROETHYLENE	20	28	18	8	30	2	3	2	24	3	138
BENZENE	20	25	30	16	23	8	5	5	4	1	137
TOLUENE	21	26	14	19	27	8	6	2	4	4	131
VOLATILE ORGANIC COMPOUNDS (VOCs)	29	37	9	12	23	4	4	1	7	3	129
ARSENIC	8	11	15	17	29	10	5	10	7	7	119
CADMIUM	2	9	12	23	22	4	9	9	3	7	100
1,1,1-TRICHLOROETHANE	16	12	16	9	15	3	4		10	1	86
COPPER	4	9	6	14	15	8	4	2	8	4	74
ZINC	1	6	11	11	13	6	7	5	5	6	71
VINYL CHLORIDE	8	15	17	9	10	4	2		3		68
XYLENE	11	8	6	9	21	5	1	1	4	1	67
CHLOROFORM	8	12	11	5	12	2	3	1	7	4	65
PHENOLS	3	11	8	14	16	5	1	3	3		64
1,1-DICHLOROETHANE	9	4	10	8	20	2			6	1	60
WASTE SOLVENTS		10	2	4	21	4	3	1	7	5	57
CYANIDES	2	1	9	7	21	2	4		1	6	53
NICKEL	5	5	8	9	9	2	3	2	1	2	46
1,1-DICHLOROETHYLENE	5	1	6	4	9	1	1	1	18		45
ETHYLBENZENE	8	8	5	5	14	3	1	1			45
METHYLENE CHLORIDE	4	10	2	7	13	1	1	2	3	2	45
PESTICIDES	4	9	1	10	4	5	2	3	5	2	45
MERCURY	3	9	8	8	5	2	1		3	3	42
POLYAROMATIC HYDROCARBONS (PAHs)	6	6	7	1	13	3	3	1	1	1	42
1,2-TRANS-DICHLOROETHYLENE	5	4	3	9	10	1	1	2	5	1	41
ORGANIC COMPOUNDS			1	9	22	5		2			40
PENTACHLOROPHENOL		2	4	7	7	5	1	5	6	3	39

Appendix A

FREQUENCY OF OCCURRENCE — EPA REGIONS

CONTAMINANTS	I	II	III	IV	V	VI	VII	VIII	IX	X	TOTAL
ACIDS	3	8	3	1	6	1	5		1	3	32
BARIUM	1		3	6	8	4	3	1		4	30
CARBON TETRACHLORIDE	4	6	2	1	4	1	3		6	3	30
CREOSOTE	2		4	6	5	6		4	2		29
METHYL ETHYL KETONE	8	1	2	2	9	1		1	2	2	28
TRICHLOROETHANE	3	3	3	2	2			2	12	1	28
WASTE OILS/SLUDGES		11	1	1	8	4		1	1		27
CHLORINATED ORGANIC COMPOUNDS		6		2	10	1			2	5	26
1,2-DICHLOROETHANE	2	1	7	2	7	1	4				24
MANGANESE	1		3	3	7	2	2	3		3	24
NAPHTHALENE		5	1	7	1	5	1	2			22
1,2-DICHLOROETHYLENE	3	3	5	1	6		1	1	1		21
DDT		3		12		4		1	1		21
DIOXINS	1	5			3	4	6	1	1		21
IRON		4		4	6	4	1	1	1		21
CHLOROBENZENE	2	1	10	1	3	1	1	2	1	2	20
PHTHALATES	2	4	5	1	3	2	1		2		17
WASTE PAINTS/LAQUERS		1		1	6	2	4			3	17
DICHLOROBENZENE	1	5	2	4	1				3		16
DICHLOROETHYLENE	2	1	4	1	4			1	3		16
ACETONE	2	1			6	1	1		3	1	15
ASBESTOS	2	2	2		2		2		5		15
1,1,1-TRICHLOROETHYLENE	2	2	3		5				2	2	14
LINDANE		2	1	10		1	1				14
BIS(2-ETHYLHEXYL)PHTHALATES	1	4	2	1	1	2	2		1		13
HALOGENATED SOLVENTS	3	4		3		1	1				13
CHLORDANE		2		8		1	1				12
AMMONIA					3						10
CAUSTICS	2	3			2	7		1			10
FURANS	1		1		4	2				1	10
INORGANIC COMPOUNDS				3	7				3		10
DIELDRIN		1		6		1	1				9

(continued)

FREQUENCY OF OCCURRENCE EPA REGIONS

CONTAMINANTS	I	II	III	IV	V	VI	VII	VIII	IX	X	TOTAL
SELENIUM		1		2	3		1	1	1		9
TOXAPHENE				7		1	1				9
ALDRIN		1		5			1	1			8
METHANE	1	3		2					2		8
METHYLISOBUTYLKETONE	2		2	1	1			1			8
PETROLEUM HYDROCARBONS		4		1				3			8
PHENANTHRENE		2		1	2	2	1		2		8
2,4-DICHLOROPHENOXYACETIC ACID (2,4-D)		1		1		2	1				7
2,4-DIMETHYLPHENOL			2	1	2	1		1	1		7
BENZO(A)PYRENE			2		2				1		7
CHLORINATED HYDROCARBONS	2		4								7
SULFURIC ACID	1			2	1		1	1		1	7
ALCOHOLS		1		1		1					6
ANTHRACENE				3	1	1	1		3		6
BERYLLIUM	1		4			1					6
DDE		1		3		1			1		6
FLUORIDE				1	1	1	2				6
PYRENE		2	1	2	1		1				6
STYRENE		2	1			2	1				6
SILVER	1	1				1	1	1			6
THORIUM			1		4		1				6
TRICHLOROPROPANE									6		6
1,2-DICHLOROPROPANE			2		1				2		5
2,4,5-TRICHLOROPHENOXYACETIC ACID (2,4,5-T)						3			2		5
BORON	1		1		1	1	1				5
FLUORANTHENE		2		2	1						5
FREON-113									5		5
HEPTACHLOR				1	1	1	2				5
METHYLPARATHION				3		1	1				5
URANIUM	1										5
ANTIMONY		1		2	2		1	1		1	4

(continued)

Appendix A   113

CONTAMINANTS	FREQUENCY OF OCCURRENCE EPA REGIONS										TOTAL
	I	II	III	IV	V	VI	VII	VIII	IX	X	
BIS(2-CHLOROETHYL)ETHER	1		1		1	1					4
COAL TARS		1		1	2						4
DDD		1		1			1		1		4
DIBROMOCHLOROPROPANE									4		4
DICHLOROETHANE					3	1					4
HYDROFLUORIC ACID					4						4
PHOSPHATES			1	2				1			4
RADIUM AND COMPOUNDS	1		1				1	1			4
SEMIVOLATILE ORGANIC COMPOUNDS		1			3						4
SULFATES (ION)					2	1	1				4
TIN				1	1	1				1	4
2-METHYLNAPHTHALENE		1		1	1						3
BENZENEHEXACHLORIDE				2	1				1		3
BENZO[A]ANTHRACENE			1	1	1						3
BENZO[J,K]FLUORENE			2	1							3
CARBON DISULFIDE			1	1		1					3
CHLORIDE (ION)					2	1					3
DI-N-BUTYL-PHTHALATES		2					1				3
DIBENZOFURAN						3					3
DIETHYLPHTHALATES		1	1					1			3
ETHYLENEDIBROMIDE							1		2		3
ETHYLENE GLYCOL				1	1	1					3
FLUORENE		1				1				1	3
HEXOCHLOROCYCLOPENTADIENE		3									3
HYDROGEN SULFIDE		1	1		1						3
NITRATES				1	1				1		3
RADIOACTIVE WASTES		1		2							3
RADON AND COMPOUNDS		3									3
TETRACHLOROETHANE				1		1			1		3
THIOCYANATES			1		2						3
TRIBROMOMETHANE		2		1							3
ACENAPTHYLENE			1				1				2

(continued)

CONTAMINANTS	FREQUENCY OF OCCURRENCE EPA REGIONS										TOTAL
	I	II	III	IV	V	VI	VII	VIII	IX	X	
ACROLEIN			1			1					2
ALUMINUM										1	2
ANILINE	1				1						2
BROMOMETHANE		1			1						2
CHLORODIBROMOMETHANE		1				1					2
CHROMIC ACID		1			1			1			2
CHRYSENE		2									2
COBALT		1				1					2
CYCLOHEXANE	1				1						2
DINITROTOLUENE			1		1						2
ETHYLACETATE	1			1							2
ETHYLPARATHION				1		1					2
ETHYLENE					1						2
ETHYLENECHLORIDE	1	1					1				2
HEXACHLOROBENZENE		1		1							2
INKS AND DYES		2									2
ISOPHORONE		1				1					2
MAGNESIUM		1					1				2
N,N-DIMETHYLFORMAMIDE	1								1		2
SODIUM HYDROXIDE				1	1						2
THIMET			1	1							2
TRICHLOROFLUOROMETHANE				1					1		2
TRICHLOROPHENOLS		2									2
TRINITROTOLUENE (TNT)					2						2
VANADIUM			2								2
1,2,4-TRICHLOROBENZENE		1		1							1
2,4,5-TRICHLOROPHENOL		1									1
ACETONITRILE		1									1
ADIPIC ACID		1									1
BENZONITRILE				1							1
BUTADIENE		1									1
CHLORINATED BENZOFLUORIDE		1									1

(continued)

Appendix A

CONTAMINANTS	FREQUENCY OF OCCURRENCE EPA REGIONS										TOTAL
	I	II	III	IV	V	VI	VII	VIII	IX	X	
DICHLORONITROETHANE				1							1
DIMETHYLPHENOL			1								1
ENDOSULFAN SULFATE		1									1
HEXACHLOROBUTADIENE		1									1
HEXYL PHTHALATE			1								1
MALEIC ACID		1									1
MERCURIC CHLORIDE						1					1
METHYL ACETATE	1										1
MOLYBDENUM								1			1
N-BUTYLACETATE									1		1
NAPALM			1								1
PENTACHLOROETHYLENE									1		1
PICOLINES					1						1
POTASSIUM CHROMATE				1							1
PYRIDINE					1						1
SEVIN		1									1
TETRACHLOROBENZENE									1		1
THALLIUM						1					1
TRICHLOROBENZENE									1		1
TRICHLOROFLUOROETHANE									1		1
TRIETHYLPHOSPHATE			1								1
TRIMELLITIC		1									1
TRISULFIDE				1							1
VINYL ACETATE		1									1

a - 1989 Guide to Superfund Sites

Appendix B

Debris/Materials Categorization

APPENDIX B

DEBRIS/MATERIALS CATEGORIZATION

<u>Textiles</u>
- Rags
- Tarpaulins
- Mattresses

<u>Glass</u>
- Bottles, ballasts
- Windows

<u>Paper</u>
- Books/magazines
- Packaging
- Cardboard/fiberboard, drums

<u>Metal</u>
- Ferrous--cast iron, tin cans, slag
- Nonferrous--aluminum, brass, copper, stainless steel
- Metal objects--drums, tanks, refrigerators, cars, gas cylinders, vans, railroad rails, transformers, capacitors, scrap metal

<u>Plastic</u>
- Battery casings
- Six-pack rings
- Buckets
- Plastic bags/sheets

<u>Rubber</u>
- Battery casings
- Tires
- Hose

Wood/Vegetation

- ○ Stumps, leaves, brush, branches
- ○ Pallets, railroad ties
- ○ Furniture

Construction Debris

- ○ Brick/concrete
- ○ Asphalt
- ○ Pipe
- ○ Plywood
- ○ Insulation--fiberglass

Soil

- ○ Clay
- ○ Topsoil
- ○ Fly ash
- ○ Sediment--stones, rocks

Sludge

- ○ Paint
- ○ Oil
- ○ Lead batteries, acid
- ○ Stillbottoms

Liquids

- ○ Oil
- ○ Water--ponds, rinsate
- ○ Acids/solvents

Abestos

- ○ Insulation
- ○ Friable debris

Appendix C

Debris/Materials Characterization for 100 Hazardous Waste Sites

DEBRIS/MATERIAL CHARACTERIZATION FOR 144 HAZARDOUS WASTE SITES[a]

Site	State	Misc.	Textiles	Glass	Metal[b]	Paper	Plastic	Rubber	Wood, Veg.	Constr. Debris	Soil	Liquid[b]	Sludge	Asbestos
Region I														
Union Chemical Co.	ME				D,T									
Cannon Engineering	MA	✓		✓	D,T	✓			✓		✓	O,W	✓	
Iron Horse Park	MA				D				✓		✓	F,S	✓	✓
Fletcher Paint	NH				D						✓	O	✓	
Ridge Ave.	NH										✓			✓
Cook's Landfill	RI			✓	D,M						✓	F,W	✓	✓
Davis Liquid Chemical	RI				D,T			✓			✓	O,W,S	✓	
Western Sand and Gravel	RI										✓	W	✓	
Region II														
Industrial Latex	NJ				D,I,M	✓			✓	✓	✓	F	F	✓
Int'l Metallurgical Services	NJ	✓		✓	D,I,M	✓				✓				
Kearny Drum Dump	NJ				D					✓	✓	F,O		
Chemical Insecticide Corp.	NY									✓	✓			
Jagger Lane	NY										✓	W		
Wide Beach	NY								✓	✓	✓	O,W		
Region III														
Delaware Sand & Gravel	DE				D					✓		W	✓	
Kane & Lombard St. Drum	MD				D		✓			✓		F	✓	
ABM-Wade	PA				D,T					✓		A,W	✓	
Ambler Asbestos Tailings	PA													✓
Boyertown Scrap	PA				D					✓	✓	F,W	✓	
Brown's Battery Breaking	PA										✓	W	✓	
Bruin Lagoon #2	PA				M		✓			✓	✓	A,H	✓	

(continued)

Site	State	Misc.	Textiles	Glass	Metal[b]	Paper	Plastic	Rubber	Wood, Veg.	Constr. Debris	Soil	Liquid[b]	Sludge	Asbestos
Maltovsky Drum	PA				D,I							F,O	✓	✓
Taylor Borough	PA				D							O,W	✓	
Tyson's Dump	PA										✓	W	✓	
Westline	PA	✓										W	✓	
Region IV														
Mowbray Engineering	AL				T						✓	O	✓	
Tower Chemical Co.	FL											W		
A.L. Taylor	KY				D				✓	✓	✓	H,L	✓	
Lee's Lane Landfill	KY				D	✓				✓	✓	H	✓	
Newport Dump Site	KY				M	✓				✓		W		
Plastifax Co.	MS					✓				✓	✓	H,V		
Aberdeen Pesticide	NC										✓	W		
American Creosote Works	TN				T						✓	W,O	✓	
Region V														
A&F Materials	IL				T						✓	O,A,W		
Belvidere Municipal Landfill	IL									✓		W	✓	
Chicago Drums	IL				D									
Gebhart Fertilizer	IL				D,I							H		
LaSalle Electrical	IL				D,I						✓	O		
U.S. Scrap	IL	✓			D							H		
Conservation Chemical	IN				D,I						✓	O,P	✓	
Envirochem	IN				T,M							W	✓	
Fell's Junkyard	IN				M						✓			
Lake Sandy Jo	IN	✓								✓	✓			
MIDCO II	IN				D,I						✓			
Midwest Plating/Kokomo	IN				D,M,I				✓		✓	W,A,H,P	✓	
Midwest Plating/Logansport	IN				D,M,I				✓			W,F,A,H	✓	

(continued)

Site	State	Misc.	Textiles	Glass	Metal[b]	Paper	Plastic	Rubber	Wood, Veg.	Constr. Debris	Soil	Liquid[c]	Sludge	Asbestos
Ninth Ave. Dump	IN				D,I	✓	✓	✓	✓	✓		O,W		
Portage Drum	IN				D							?		
Tyler Street Drums	IN				D								✓	
Carter Salvage	MI	✓			M						✓	O		
Duell & Gardner	MI				D						✓		✓	
Forest Waste	MI				D				✓		✓		✓	
G & H Landfill	MI								✓		✓	O,S	✓	
Liquid Disposal	MI	✓		✓	M,I				✓		✓	F,W		
PBM Enterprises	MI										✓	H,W		
Rasmussen Dump	MI				D						✓			
Saginaw Paint	MI				D						✓			
Verona Well Field	MI				T		✓				✓	W		
Barney Rumple	MN				D						✓			
Union Scrap	MN	✓						✓			✓	?		
American Steel Drum	OH	✓			D	✓					✓	W		
Dayton Tire & Rubber	OH				D,M,I						✓	W		
Industrial Excess Landfill	OH	✓				✓			✓	✓	✓			
Republic Hose	OH				D,M					✓	✓	O		
Better Brite Plating	WI				D						✓	W		
Lee's Farm	WI						✓				✓	W		
Try Chem	WI	✓			D,I							W		
Region VI														
Allen Transformer	AR				D		✓			✓	✓	W	✓	
Cleve Reber	LA				D		✓		✓		✓	W		
Crystal City Airport	TX								✓		✓			
Geneva Industries	TX									✓	✓	W,O		
Motco Site	TX				T							H,W	✓	

(continued)

Appendix C   123

Site	State	Misc.	Textiles	Glass	Metal[b]	Paper	Plastic	Rubber	Wood, Veg.	Constr. Debris	Soil	Liquid[b]	Sludge	Asbestos
Passes Chemical Co.	TX												✓	
Stevco Site	TX											O,W	✓	
Triangle Chemical	TX	✓			D						✓	A,F,W		
Region VII														
B&B Salvage Co.	MO				M						✓			
Broadway Salvage Oil #1	MO								✓		✓			
Minker Cul-de-Sac/Area	MO	✓			M				✓		✓			
Posch Foundry	MO				M						✓	A,H		✓
Quail Run	MO									✓	✓	O		
Solid State Circuits	MO									✓	✓	W		
Region VIII														
B&C Metals	CO									✓				
Eagle Mine	CO											O	✓	
Gene Murren Property	CO									✓	✓	A	✓	
Mestos Well	CO										✓		✓	
PDC Spas	CO										✓	F,H		
Woodbury Chemical	CO									✓	✓		✓	
Burlington Northern RR	MT				T				✓		✓	O	✓	
Montana Pole NPL	MT										✓	W,O		
Arlington Spill Site	WY													
Region IX														
McColl Site	CA										✓			
Region X														
Alaska Battery Enterprises	AK						✓	✓						
Ohlsen Mountain	AK				M					✓	✓	W	✓	
ArrComm Corp.	ID				T						✓	O	✓	
Bunker Hill	ID										✓	O,H	✓	✓

(continued)

Site	State	Misc.	Textiles	Glass	Metal[b]	Paper	Plastic	Rubber	Wood, Veg.	Constr. Debris	Soil	Liquid[b]	Sludge	Asbestos
Pacific Hide & Fur	ID			✓	D,M						✓			
Northwest Dust Control	OR				D,I						✓	O,W	✓	
American Crossarm & Conduit	WA								✓		✓	O	✓	
Northwest Transformer	WA				M				✓	✓	✓	O		
William & Son Transformer	WA				T						✓	O		✓

[a] This list contains what was mentioned as being on site, not what actually may have been present.

[b]
A = Acids
D = Drums
F = Flammable
H = Hazardous liquid
M = Metal objects
O = Oil
S = Solvents
T = Tanks, vats
W = Water

Appendix D

Equipment Used at 100 Hazardous Waste Sites

REGION I

	Union Chemical ME	Cannon Engineering, MA	Iron Horse Park, MA	Fletcher Paint, NH	Ridge Ave., NH	Cooks Landfill, RI	Davis Liquid Chemical, RI	Western Sand & Gravel, RI
Backhoe/excavator	✓	✓	✓	✓		✓		✓
Front-end loader	✓		✓				✓	
Bulldozer	✓		✓	✓	✓			✓
Lowboy	✓							
Dump truck				✓		✓		
Tractor (OTR)	✓	✓			✓			
Landfill compactor								
Grader					✓			
Crane					✓			
Forklift								
Skid-steer loader				✓		✓	✓	✓
Diaphragm pump	✓							✓
Trash pump								
Submersible pump								
Barrel pump								✓
Vacuum unit								✓
Vacuum truck	✓							✓
Pressure washer/laser								
Crusher (drum/debris)				✓				

(continued)

REGION I (continued)

	Union Chemical MN	Cannon Engineering, MA	Iron Horse Park, MA	Fletcher Paint, NH	Ridge Ave., NH	Cooks Landfill, RI	Davis Liquid Chemical, RI	Western Sand & Gravel, RI
Shredder (tire, drum)								
Vibrating screen								
Drum cart	✓						✓	
Drum punch							✓	
Drum grappler	✓	✓						
Holding/bulking tanks								
Generator	✓						✓	✓
Air compressor								✓
Air hammer								
Chain saw			✓	✓	✓	✓		
Cutting torch					✓	✓		
Hand tools	✓						✓	
Non-sparking tools	✓							
Hoe ram/pile driver								
Pug mill								
Rolloff box		✓		✓				
Dragline								
Conveyor								
Portable building								

Appendix D   127

REGION II

	Industrial Latex, NJ	International Metallurgical Services, NJ	Kearny Drum Dump, NJ	Chemical Insecticide Corp., NY	Jagger Lane, NY	Wide Beach, NY
Backhoe/excavator	✓	✓	✓	✓	✓	
Front-end loader	✓			✓	✓	
Bulldozer		✓		✓		✓
Lowboy			✓			
Dump truck						
Tractor (OTR)			✓			
Landfill compactor						
Grader						✓
Crane	✓					
Forklift	✓	✓				
Skid-steer loader	✓					
Diaphragm pump	✓					✓
Trash pump						
Submersible pump	✓		✓			
Barrel pump						
Vacuum unit		✓				✓
Vacuum truck		✓				
Pressure washer/laser	✓			✓		
Crusher (drum/debris)						

(continued)

REGION II (continued)

	Industrial Latex, NJ	International Metallurgical Services, NJ	Kearny Drum Dump, NJ	Chemical Insecticide Corp., NY	Jagger Lane, NY	Wide Beach, NY
Shredder (tire, drum)	✓					
Vibrating screen						
Drum cart			✓			
Drum punch						
Drum grappler	✓		✓			
Holding/bulking tanks	✓					
Generator	✓		✓			
Air compressor						
Air hammer					✓	✓
Chain saw	✓					
Cutting torch	✓		✓	✓	✓	
Hand tools						
Non-sparking tools						
Hoe ram/pile driver						
Rolloff box	✓	✓				
Dragline						
Conveyor						
Portable building						

Appendix D   129

REGION III

	Delaware Sand & Gravel, DE	Kane & Lombard Drum, MD	ABM Wade, PA	Ambler Asbestos Tailings, PA	Boyertown Scrap Metal, PA	Brown's Battery Breaking, PA	Bruin Lagoon, PA	Malitovsky Drum Co., PA	Taylor Borough, PA	Tyson's Dump, PA	Westline, PA
Backhoe/excavator	✓	✓	✓	✓		✓	✓	✓	✓	✓	✓
Front-end loader	✓	✓	✓	✓		✓	✓	✓	✓		
Bulldozer		✓		✓		✓	✓	✓	✓		
Lowboy	✓										
Dump truck	✓					✓				✓	
Tractor (OTR)	✓	✓		✓							✓
Landfill compactor											
Grader								✓			
Crane											
Forklift				✓							
Skid-steer loader			✓					✓			
Diaphragm pump		✓					✓	✓			
Trash pump											
Submersible pump										✓	
Barrel pump											
Vacuum unit								✓			
Vacuum truck								✓			
Pressure washer/laser			✓	✓							
Crusher (drum/debris)											

(continued)

REGION III (continued)

	Delaware Sand & Gravel, DE	Kane & Lombard Drum, MD	ABM Wade, PA	Ambler Asbestos Tailings, PA	Boyertown Scrap Metal, PA	Brown's Battery Breaking, PA	Bruin Lagoon, PA	Malitovsky Drum Co., PA	Taylor Borough, PA	Tyson's Dump, PA	Westline, PA
Shredder (tire drum)	✓	✓									
Vibrating screen	✓										
Drum cart				✓							
Drum punch				✓				✓			
Drum grappler	✓	✓						✓			
Holding/bulking tanks		✓						✓		✓	
Generator	✓	✓			✓				✓		✓
Air compressor		✓							✓		
Air hammer							✓				
Chain saw		✓									
Cutting torch				✓							
Hand tools			✓	✓	✓			✓			✓
Non-sparking tools					✓			✓			
Hoe ram/pile driver				✓							
Pug mill			✓		✓						
Rolloff box											
Dragline											
Conveyor											
Portable building											

Appendix D 131

132 Materials-Handling Technologies Used at Hazardous Waste Sites

REGION IV

	Mowbray Engineering, AL	Tower Chemical, FL	A.L. Taylor, KY	Lee's Lane Landfill, KY	Newport Dump, KY	Plastifax Co., MS	Aberdeen Pesticide, NC	American Creosote, TN
Backhoe/excavator	✓	✓	✓	✓	✓		✓	✓
Front-end loader	✓	✓	✓	✓	✓		✓	✓
Bulldozer	✓		✓	✓	✓	✓		
Lowboy	✓	✓	✓	✓	✓		✓	
Dump truck	✓		✓	✓	✓		✓	
Tractor (OTR)	✓	✓	✓	✓				
Landfill compactor								
Grader					✓			
Crane								
Forklift								
Skid-steer loader	✓							
Diaphragm pump				✓			✓	
Trash pump	✓	✓	✓		✓			
Submersible pump								
Barrel pump	✓					✓		
Vacuum unit						✓		
Vacuum truck	✓				✓			
Pressure washer/laser	✓	✓	✓	✓	✓			
Crusher (drum/debris)	✓		✓	✓	✓			

(continued)

REGION IV (continued)

	Mowbray Engineering, AL	Tower Chemical, FL	A.L. Taylor, KY	Lee's Lane Landfill, KY	Newport Dump, KY	Plastifax Co., MS	Aberdeen Pesticide, NC	American Creosote, TN
Shredder (tire, drum)					✓		✓	
Vibrating screen							✓	
Drum cart						✓		
Drum punch								
Drum grappler							✓	
Holding/bulking tanks	✓	✓	✓	✓				
Generator	✓		✓			✓		
Air compressor	✓		✓	✓	✓			
Air hammer				✓	✓			
Chain saw				✓				
Cutting torch	✓							
Hand tools						✓		
Non-sparking tools						✓		
Hoe ram/pile driver								
Pug mill								
Rolloff box								
Dragline						✓		
Conveyor							✓	
Portable building								

REGION V

	Conservation Chemical, IN	MIDCO II, IN	Midwest Plating-Kokomo, IN	Midwest Plating-Logansport, IN	Lake Sandy Jo, IN	9th Ave Dump, IN	Envirochem-Zionsville, IN
Backhoe/excavator	✓	✓	✓	✓			✓
Front-end loader	✓	✓	✓	✓		✓	✓
Bulldozer					✓	✓	✓
Lowboy	✓	✓	✓	✓			
Dump truck							
Tractor (OTR)		✓	✓				
Landfill compactor							
Grader				✓			
Crane				✓			
Forklift			✓	✓			
Skid-steer loader			✓	✓			
Diaphragm pump	✓	✓	✓	✓		✓	✓
Trash pump	✓	✓					
Submersible pump	✓	✓	✓	✓			
Barrel pump	✓						
Vacuum unit		✓	✓	✓			
Vacuum truck	✓			✓			
Pressure washer/laser		✓	✓				✓
Crusher (drum/debris)							

(continued)

REGION V (continued)

	Conservation Chemical, IN	MIDCO II, IN	Midwest Plating-Kokomo, IN	Midwest Plating-Logansport, IN	Lake Sandy Jo, IN	9th Ave Dump, IN	Envirochem-Zionsville, IN
Shredder (tire, drum)		✓					
Vibrating screen							
Drum cart	✓						
Drum punch		✓					
Drum grappler	✓	✓					
Holding/bulking tanks	✓	✓	✓	✓			✓
Generator	✓	✓	✓	✓			✓
Air compressor	✓	✓		✓			✓
Air hammer	✓						✓
Chain saw					✓		
Cutting torch		✓	✓				✓
Hand tools	✓	✓					✓
Non-sparking tools	✓	✓	✓				
Hoe ram/pile driver							✓
Pug mill							
Rolloff box							
Dragline							
Conveyor							
Portable building							

(continued)

REGION V (continued)

	Tyler St. Drum, IN	Portage Drums, IN	Fell Junkyard, IN	A&F Greenup, IL	Belvidere Landfill, IL	Chicago Drums, IL	Gebhart Fertilizer, IL	U.S. Scrap Chicago, IL	Lasalle Electric, IL	Carter Salvage, MI
Backhoe/excavator			✓					✓		✓
Front-end loader			✓		✓		✓	✓	✓	✓
Bulldozer										✓
Lowboy	✓	✓		✓					✓	
Dump truck										
Tractor (OTR)					✓		✓		✓	
Landfill compactor								✓		
Grader										
Crane										✓
Forklift	✓									✓
Skid-steer loader		✓	✓			✓	✓			
Diaphragm pump				✓		✓				
Trash pump									✓	✓
Submersible pump										✓
Barrel pump										
Vacuum unit							✓			
Vacuum truck								✓		✓
Pressure washer/laser								✓		✓
Crusher (drum/debris)	✓	✓								✓

(continued)

REGION V (continued)

	Tyler St. Drum, IN	Portage Drums, IN	Fell Junkyard, IN	A&F Greenup, IL	Belvidere Landfill, IL	Chicago Drums, IL	Gebhart Fertilizer, IL	U.S. Scrap Chicago, IL	Lasalle Electric, IL	Carter Salvage, MI
Shredder (tire, drum)										✓
Vibrating screen										
Drum cart		✓							✓	✓
Drum punch	✓							✓		✓
Drum grappler	✓				✓			✓		
Holding/bulking tanks					✓			✓		
Generator					✓			✓	✓	✓
Air compressor				✓		✓				
Air hammer								✓		
Chain saw										
Cutting torch										✓
Hand tools	✓		✓					✓	✓	
Non-sparking tools								✓		
Hoe ram/pile driver										
Pug mill										
Rolloff box										
Dragline Conveyor										
Portable building										

(continued)

REGION V (continued)

	Liquid Disposal, MI	G & H Landfill, MI	PBM Enterprises, MI	Forest Waste, MI	Saginaw Paint, MI	Verona Well Field, MI	Rasmussen Dump, MI	Duell & Gardner, MI	Barney Rumple Junkyard, MN	Union Scrap Iron & Metal, MN
Backhoe/excavator	✓	✓	✓		✓	✓	✓	✓	✓	✓
Front-end loader	✓	✓	✓		✓	✓	✓	✓	✓	✓
Bulldozer		✓		✓		✓	✓	✓	✓	
Lowboy	✓		✓		✓					
Dump truck		✓	✓							
Tractor (OTR)		✓								
Landfill compactor										
Grader		✓								
Crane	✓					✓				
Forklift			✓		✓					
Skid-steer loader			✓					✓		
Diaphragm pump	✓	✓	✓		✓					
Trash pump		✓				✓				
Submersible pump	✓	✓	✓			✓				
Barrel pump										
Vacuum unit					✓		✓			
Vacuum truck	✓	✓	✓				✓		✓	
Pressure washer/laser										
Crusher (drum/debris)										

(continued)

REGION V (continued)

	Liquid Disposal, MI	G & H Landfill, MI	PBM Enterprises, MI	Forest Waste, MI	Saginaw Paint, MI	Verona Well Field, MI	Rasmussen Dump, MI	Duell & Gardner, MI	Barney Rumple Junkyard, MN	Union Scrap Iron & Metal, MN
Shredder (tire, drum)							✓			
Vibrating screen	✓									
Drum cart	✓									
Drum punch	✓									
Drum grappler			✓							
Holding/bulking tanks	✓						✓			
Generator	✓		✓	✓						
Air compressor	✓		✓			✓			✓	✓
Air hammer									✓	
Chain saw			✓	✓						
Cutting torch	✓				✓	✓				
Hand tools			✓		✓	✓			✓	✓
Non-sparking tools									✓	
Hoe ram/pile driver										
Pug mill							✓			
Rolloff box	✓									
Dragline										
Conveyor										
Portable building										

(continued)

REGION V (continued)

	American Steel Drum, OH	Dayton Tire & Rubber, OH	Industrial Excess Landfill, OH	Republic Hose, OH	Better Brite Plating, WI	Lee's Farm, WI	Try Chem Milwaukee, WI
Backhoe/excavation	✓	✓	✓	✓		✓	
Front-end loader	✓	✓	✓			✓	
Bulldozer			✓			✓	
Lowboy	✓	✓	✓	✓		✓	✓
Dump truck		✓	✓				
Tractor (OTR)	✓		✓	✓			
Landfill compactor							
Grader		✓					
Crane		✓	✓	✓		✓	
Forklift							
Skid-steer loader	✓	✓			✓	✓	
Diaphragm pump	✓	✓		✓	✓	✓	✓
Trash pump	✓	✓	✓	✓		✓	
Submersible pump							✓
Barrel pump	✓						
Vacuum unit							
Vacuum truck	✓	✓		✓			
Pressure washer/laser	✓		✓			✓	
Crusher (drum/debris)						✓	

(continued)

REGION V (continued)

Equipment	American Steel Drum, OH	Dayton Tire & Rubber, OH	Industrial Excess Landfill, OH	Republic Hose, OH	Better Brite Plating, WI	Lee's Farm, WI	Try Chem Milwaukee, WI
Shredder (tire, drum)							
Vibrating screen							
Drum cart		✓					
Drum punch			✓				
Drum grappler				✓			
Holding/bulking tanks	✓			✓			
Generator	✓	✓	✓	✓		✓	
Air compressor	✓	✓	✓	✓	✓	✓	
Air hammer	✓		✓				
Chain saw			✓	✓		✓	
Cutting torch	✓		✓		✓		
Hand tools	✓			✓		✓	
Non-sparking tools							
Hoe ram/pile driver							
Pug mill							
Rolloff box	✓						
Dragline							
Conveyor						✓	
Portable building							

142 Materials-Handling Technologies Used at Hazardous Waste Sites

REGION VI

	Allen Transformer, AR	Cleve Reber, LA	Crystal City Airport, TX	Geneva Industries, TX	MOTCO, TX	Pesses Chemical, TX	Stewco, TX	Triangle Chemical, TX
Backhoe/excavator	✓	✓	✓	✓		✓	✓	✓
Front-end loader	✓	✓	✓			✓	✓	✓
Bulldozer	✓	✓	✓	✓				
Lowboy	✓							
Dump truck	✓	✓	✓			✓	✓	
Tractor (OTR)								
Landfill compactor								
Grader			✓					
Crane	✓			✓				✓
Forklift	✓			✓				
Skid-steer loader	✓							
Diaphragm pump				✓	✓	✓		
Trash pump								
Submersible pump				✓			✓	
Barrel pump								
Vacuum unit								
Vacuum truck								
Pressure washer/laser	✓		✓			✓		✓
Crusher (drum/debris)								

(continued)

REGION VI (continued)

	Allen Transformer, AR	Cleve Reber, LA	Crystal City Airport, TX	Geneva Industries, TX	MOTCO, TX	Pesses Chemical, TX	Stewco, TX	Triangle Chemical, TX
Shredder (tire, drum)								
Vibrating screen				✓				
Drum cart								
Drum punch								
Drum grappler								✓
Holding/bulking tanks	✓			✓	✓			
Generator				✓	✓			
Air compressor								
Air hammer								
Chain saw				✓				
Cutting torch								
Hand tools	✓			✓		✓	✓	
Non-sparking tools								
Hoe ram/pile driver								
Pug mill								
Rolloff box	✓	✓		✓		✓	✓	✓
Dragline								
Conveyor								
Portable building								

REGION VII

	B&B Salvage, MO	Broadway Salvage Oil #1, MO	Minker Area, Cul de Sac, MO	Posch Foundry, MO	Quail Run, MO	Solid State Circuits, MO
Backhoe/excavator	✓	✓	✓	✓	✓	✓
Front-end loader		✓	✓		✓	✓
Bulldozer		✓	✓			
Lowboy						
Dump truck	✓	✓	✓	✓	✓	✓
Tractor (OTR)						
Landfill compactor						
Grader			✓			
Crane			✓	✓	✓	
Forklift	✓					✓
Skid-steer loader						
Diaphragm pump						
Trash pump						
Submersible pump						
Barrel pump						
Vacuum unit	✓		✓		✓	
Vacuum truck						
Pressure washer/laser	✓		✓		✓	
Crusher (drum/debris)						

(continued)

REGION VII (continued)

	B&B Salvage, MO	Broadway Salvage Oil #1, MO	Minker Area, Cul de Sac, MO	Posch Foundry, MO	Quail Run, MO	Solid State Circuits, MO
Shredder (tire, drum)						
Vibrating screen						
Drum cart						
Drum punch						
Drum grappler	✓					
Holding/bulking tanks						
Generator			✓		✓	
Air compressor					✓	✓
Air hammer			✓			
Chain saw						✓
Cutting torch	✓					
Hand tools						
Non-sparking tools						
Hoe ram/pile driver						
Pug mill						
Rolloff box			✓		✓	
Dragline		✓				
Conveyor						
Portable building						

REGION VIII

	B&C Metals, CO	Eagle Mine, CO	Gene Mureen, CO	Mestas Well, CO	PDC Spas, CO	Woodbury Chemical, CO	Burlington Northern, MT	Montana Pole, MT	Arlington Spill Site, WY
Backhoe/excavator				✓	✓		✓	✓	✓
Front-end loader			✓			✓	✓	✓	
Bulldozer		✓				✓		✓	
Lowboy					✓	✓			
Dump truck		✓	✓			✓	✓		
Tractor (OTR)	✓								
Landfill compactor									
Grader								✓	
Crane								✓	
Forklift									
Skid-steer loader							✓		
Diaphragm pump									
Trash pump									
Submersible pump	✓							✓	
Barrel pump									
Vacuum unit									
Vacuum truck									
Pressure washer/laser								✓	
Crusher (drum/debris)									

(continued)

REGION VIII (continued)

	B&C Metals, CO	Eagle Mine, CO	Gene Mureen, CO	Mestas Well, CO	PDC Spas, CO	Woodbury Chemical, CO	Burlington Northern, MT	Montana Pole, MT	Arlington Spill Site, WY
Shredder (tire, drum)									
Vibrating screen									
Drum cart									
Drum punch									
Drum grappler									
Holding/bulking tanks							✓		
Generator	✓								
Air compressor									
Air hammer									
Chain saw									
Cutting torch	✓								
Hand tools	✓				✓		✓		
Non-sparking tools									
Hoe ram/pile driver									
Pug mill									
Rolloff box					✓		✓		✓
Dragline									
Conveyor									
Portable building									

Appendix D 147

REGION IX

	McColl, CA
Backhoe/excavator	✓
Front-end loader	
Bulldozer	
Lowboy	
Dump truck	
Tractor (OTR)	
Landfill compactor	
Grader	
Crane	
Forklift	✓
Skid-steer loader	✓
Diaphragm pump	
Trash pump	
Submersible pump	
Barrel pump	
Vacuum unit	
Vacuum truck	
Pressure washer/laser	✓
Crusher (drum/debris)	✓
Shredder (tire, drum)	
Vibrating screen	✓
Drum cart	
Drum punch	
Drum grappler	
Holding/bulking tanks	
Generator	✓
Air compressor	
Air hammer	

(continued)

REGION IX (continued)

	McColl, CA
Chain saw	
Cutting torch	
Hand tools	✓
Non-sparking tools	
Hoe ram/pile driver	
Pug mill	✓
Rolloff box	✓
Dragline	
Conveyor	
Portable building	✓

REGION X

	Alaska Battery, AK	Ohlson Mountain, AK	Arrcomm Corp., ID	Bunker Hill, ID	Pacific Hide & Fur, ID	NW Dust Control Facilities, OR	American Crossarm & Conduit, WA	Northwest Transformer, WA	William & Son Transf. & Salv., WA
Backhoe/excavator	✓		✓	✓	✓				✓
Front-end loader	✓		✓	✓	✓				✓
Bulldozer	✓				✓			✓	
Lowboy	✓							✓	✓
Dump truck				✓					
Tractor (OTR)				✓					
Landfill compactor									
Grader				✓					
Crane			✓		✓	✓		✓	
Forklift		✓	✓				✓		
Skid-steer loader		✓	✓	✓		✓			
Diaphragm pump			✓			✓		✓	
Trash pump			✓						
Submersible pump					✓				
Barrel pump								✓	
Vacuum unit									
Vacuum truck			✓			✓	✓	✓	
Pressure washer/laser			✓				✓	✓	✓
Crusher (drum/debris)									

(continued)

REGION X (continued)

	Alaska Battery, AK	Ohlson Mountain, AK	Arrcom Corp., ID	Bunker Hill, ID	Pacific Hide & Fur, ID	NW Dust Control Facilities, OR	American Crossarm & Conduit, WA	Northwest Transformer, WA	William & Son Transf. & Salv., WA
Shredder (tree, drum)									
Vibrating screen									
Drum cart		✓			✓				
Drum punch									
Drum grappler			✓						
Holding/bulking tanks									
Generator		✓	✓		✓				✓
Air compressor			✓					✓	✓
Air hammer		✓	✓					✓	✓
Chain saw		✓	✓				✓		
Cutting torch		✓	✓			✓			
Hand tools				✓			✓		
Non-sparking tools									
Hoe ram/pile driver						✓			✓
Pug mill									
Rolloff box	✓								
Dragline									
Conveyor									

(continued)

Appendix E

Debris/Material-Handling Overview for 67 Hazardous Waste Sites

Appendix E   153

Site name	EPA Region	Major contaminants	Principal debris/ materials handled	Debris/materials-handling procedure/equipment
Union Chemical, ME	I	Flammable solvents	4000 gal sludge; >5000 gal flammable liquids; >200 drums	Excavation of sludge; vacuum trucks used for liquids. Drums crushed with loader/backhoe.
Cannon Engineering, MA	I	Shock-sensitive materials, PCBs, Metals	475 5-gal pails	Pails were opened remotely by a skid-steer loader with a drum attachment; contents bulked for removal.
*Iron Horse Park, MA	I	Asbestos	6000 yd³ of asbestos in exposed piles	Heavy equipment was used to grade the piles and cap them; heavy vegetation required site preparation and widening of access roads.
Fletcher Paint, NH	I	Volatile organics, PCBs	Several hundred drums	Compatible wastes were bulked and overpacked for offsite disposal.
Ridge Avenue, NH	I	Asbestos	Asbestos waste pile 100 ft x 50 ft x 30 ft	Site was covered and stabilized with gravel, sand, and loam by using heavy equipment, including a gradall.
Cooks Landfill, RI	I	Asbestos, organic solvents	172 drums; 7 yd³ asbestos; glass debris	Asbestos was boxed for disposal. Drums were shipped for offsite disposal.
Davis Liquid Chemical, RI	I	1,1,1-Trichloroethane, tetrachloroethylene, benzene	>200 drums	Drums staged and crushed with backhoe; site preparation included construction of an access road and a clay pad for drum-crushing operations.
*Western Sand & Gravel, RI	I	Volatile organics	42,000 gal hazardous liquids; 297 drums of solidified material; 29 drums crushed	Liquids from storage lagoon were pumped out by vacuum truck; dike was constructed around solidified sludge that was not drummed; loader was used for drum crushing.
*Industrial Latex, NJ	II	PCBs, shock-sensitive liquid, flammable organic liquids	4381 gal of flammable organic liquids; 113,050 gal of PCB-containing liquids	Drum shredder was used to crush >900 drums; shredder was loaded with a backhoe and drum grappler; 125-ton crane was used to hoist 22 USTs onto platform for draining; water lasers were used to cut up potentially explosive USTs.
*International Metallurigal Services, NJ	II	Heavy metals, potassium cyanide, acids, shock-sensitive liquids	Three 30-yd³ rolloffs of debris; 480 drums of spent photographic film	103 drums crushed with a backhoe. Debris was loaded into rolloffs and landfilled. Spent film was loaded into drums by using shovels and a high-powered vacuum.

(continued)

154 Materials-Handling Technologies Used at Hazardous Waste Sites

Site name	EPA Region	Major contaminants	Principal debris/ materials handled	Debris/materials-handling procedure/equipment
Kearny Drum Dump, NJ	II	Lead, chromium, toluene, methylene chloride	290 leaking or deteriorated drums; top 3 in. of soil on the 1-acre site scraped off.	32 drums crushed with backhoe.
Chemical Insecticide Corp., NY	II	Pesticides 2,3,7,8-TCDD, arsenic	Leachate was collected from a parking lot; 250 feet of trench cleared of debris.	Trackhoe was used to clear the trench and to reinforce the berm.
Jagger Lane, NY	II	Trichloroethylene, tetrachloroethylene, toluene	7192 feet of water main installation and tie-ins for 78 homes with contaminated wells.	A backhoe and loader were used in this pipe laying and water hookup construction job.
Wide Beach, NY	II	PCBs	20,000 yd^3 of roads and shoulders to be paved, graded, and compacted to mitigate PCB contamination.	A gradall and bulldozer were used to shape the existing PCB-contaminated soil for application of an emulsion seal and 4 in. of asphaltic paving and surface seal.
Delaware Sand & Gravel Landfill, DE	III	Benzene, heavy metals, TCE, PCBs	975 empty drums; 575 drums containing a variety of contaminants	A backhoe with drum grappler attachment was used to load liquids plus fly ash into a compatibility chamber for solidification. A drum shredder was also used for empty drums.
Kane and Lombard Streets Drum Site, MD	III	Paint waste, volatile organics, heavy metals, toluene	120 yd^3 of shredded drums; 55 overpacked drums	1115 RCRA "empty" drums (<2 inches of material) were shredded in a clay-lined cell covered with a polyethylene liner to contain residual spillage. Equipment problems caused the shredder to be shut down for several days while parts were located.
ABM-Wade, PA	III	Heavy metals, PCBs, cyanides acids	Approximately 5000 full or partially full drums; 400 yd^3 of contaminated soils; 1300 yd^3 contaminated debris.	The drums containing hazardous materials were mixed in with scattered and deteriorated building debris, which made the removal and staging of the drums (with a backhoe and loader) very slow.
*Ambler Asbestos Tailings Pile, PA	III	Asbestos	2 exposed asbestos tailings piles --approximately 600,000 yd^3 of asbestos-laden material.	The piles were covered with soil and compacted by using heavy equipment. Steepness of the piles required a large bulldozer to winch a smaller bulldozer up and down the slope to compact the soil.

(continued)

Site name	EPA Region	Major contaminants	Principal debris/ materials handled	Debris/materials-handling procedure/equipment
Browns Battery Breaking, PA	III	Lead	72,000 yd³ of contaminated soil/ casings; 20,000 yd³ of backfill/ cover material.	Bulldozer, backhoe, and dumptrucks were used to break up frozen lead-contaminated soil and to transport it to an onsite disposal area. Disposal area measured 600 ft x 230 ft x 7 ft and was covered with clay and topsoil. Numerous instances of equipment failure due to metal fatigue from frozen soil.
*Bruin Lagoon, PA	III	Acids, hydrogen sulfide, sulfur dioxide	Backfill material to fill in old lagoon; solidified sludge in the lagoon.	A bulldozer with a ripper blade was used to break up the solidified sludge in the lagoon. Monitoring wells were installed to release trapped gases.
Hutchinson Mine, PA	III	PCBs	251 tons of contaminated soil; 495 gallons of contaminated diesel fluid.	Soil was loaded into rolloff boxes for disposal ("drum dri" fixative added because of high water content).
Malitovsky Drum Co., PA	III	Volatile organics, heavy metals, PCBs, asbestos	1052 drums; 1282 tons soil; 24,235 gallons aqueous liquids; 7 tanks	A backhoe and front-end loader were used for soil excavation. Skid-steer loader was used to stage drums; a drum shredder was brought on site to shred drums.
Taylor Borough, PA	III	PAHs, phthalate acid esters	None	Installed fences and warning signs.
Tyson's Dump, PA	III	Toluene, xylene, 1,2,3-trichloro- propane	Contaminated leachate was treated and discharged on site.	800-ft leachate collection pipeline and gravel bed were installed by using tracked excavator.
Westline, PA	III	Phenol, 2,4-dimethylphenol	Sludge in a lagoon.	Sludge was removed from the lagoon with a backhoe; fill was trucked in to restore original site characteristics.
Tower Chemical, FL	IV	DDT, DDE, DDD, xylenes	70 drums; excavated soil; wastewater from the lagoon.	Burn/burial area of 2275 ft² was excavated to a depth of 8 ft with backhoe/excavators and loaders. Contaminated wastewater was pumped from lagoon.

(continued)

156 Materials-Handling Technologies Used at Hazardous Waste Sites

Site name	EPA Region	Major contaminants	Principal debris/ materials handled	Debris/materials-handling procedure/equipment
*A. L. Taylor, KY	IV	Methylene chloride, phthalates, xylene, heavy metals, chromium and aliphatic acids	Excavated soil; buried drums; vegetation; surface water.	Drum crushing was conducted with a backhoe and loader; a retention pond was dug to contain runoff water; site was cleared of vegetation, leveled, and compacted.
Newport Steel, KY	IV	Methylene chloride, ethylbenzene, lead, 1,2-dichloroethane, benzene, toluene	Excavated soil	Site work involved excavation of soil for the installation of a leachate collection system and subsequent backfilling and grading of the area.
Plastifax Co., MS	IV	Chlorinated paraffin, sulfuric/nitric acid, caustic soda, phosphoric acid	Contaminated water in marshy area (4000 gal); removed an additional foot of soil from marsh.	Standing liquid was pumped into a holding tank. A dragline was used to remove top foot of soil. A lime slurry was sprayed over the area because of low pH.
*Aberdeen Pesticide, NC	IV	Pesticides	22,000 yd^3 of soil; crushed drums.	Trackhoe and front-end loader were used to stage soil and load it into a hopper/conveyor for transport to a power screening apparatus prior to incineration. Rejected debris from screening was shredded.
Gebhart Fetilizer Co., IL	V	DDT, 2,4-D, N-methyl-N-nitrosoethanamine, Naphthalene	Removed 155 tons of dry pesticides, 300 tons of solid fertilizer, and 3800 gal of liquid fertilizer.	Loose solid pesticides and herbicides were bagged, primarily by hand; faulty transfer pump required that liquid drummed material be hoisted to rail car by skid-steer loader and dumped by hand.
*MIDCO II, IN	V	PCBs, cadmium, lead, cyanide, methylene chloride, 1,1,1-trichloroethane, ethylbenzene	Removed more than 10,000 tons of contaminated soil/solids and 5000 gallons at contaminated liquids (including PCB waste).	Backhoe was used to excavate sludge pits; cutting torches were used to cut up aboveground tanks; approximately 65,000 drums were staged, separated, and shredded using trackhoes, grapplers, and drum shredder.
Midwest Plating & Chemical Corp. - Kokomo, IN	V	Trichloroethene, perchloroethene, copper and potassium cyanides, acids	Large amount of debris including process vats, scrap metal, and wood debris.	The debris was decontaminated using a 12.5% solution of sodium hypochlorite with a water laser, vacuum truck, and 12,000 gal portable pool. Metal debris was recovered, wood was landfilled.

(continued)

Site name	EPA Region	Major contaminants	Principal debris/materials handled	Debris/materials-handling procedure/equipment
*Midwest Plating & Chemical Corp. - Logansport, IN	V	Tetrachloroethylene, chromic acid, hydrogen cyanide gas, chromium, cadmium	More than 100,000 gal of liquid waste was treated; plating sludge was solidified with kiln dust.	Drums were staged and crushed with a forklift and loader; a vacuum truck was used to pick up sludge from underground chamber; a cutdown 55-gal drum and winch were used to haul solids out of the underground chamber that could not be picked up by vacuum truck.
*G&H Landfill, MI	V	PCBs, ethylbenzene, cyanide, mercury, lead phenols	High-solids-content liquids	A hose nozzle connected to a trash pump was attached to a trackhoe to move and collect oils from an interceptor/collector trench before collection by a vacuum track; extensive site preparation was necessary prior to work activities because the site was located on a wetland.
Liquid Disposal, Inc., MI	V	Acids, bases, paint waste, solvents, liquid & solid isocyanates	More than 800 drums were crushed (and disposed of offsite); approximately 10,000 gal of waste liquid was disposed of offsite.	Drum crusher was used for crushing the drums; liquid isocyanates packed in 15-gal plastic cubes for transporting to incinerator; most liquid wastes pumped to bulking tanker.
*PBM Enterprises, MI	V	Cyanide	1280 yd³ of debris/film chips; approximately 200,000 gal of liquid wastes and wastewater.	Onsite treatment of waste was performed using sodium hypochlorite; reaction vessels were formed from rolloff boxes lined with PVC; a concrete pad was constructed and a sump dug to collect runoff/spills.
*Aeroquip/Republic Hose, OH	V	PCB oils	Approximately 500 drums; 14 transformer casings; 1 capacitor; 285 tons PCB-contaminated soil/solids	Oil soaked ash in basement was removed by chipping into small chunks (picks and shovels) and using a vacuum unit, backhoe was used to excavate soil, transformer casings lowered from second story of building using chains and backhoe.
Lee's Farm, WI	V	Lead	Approximately 15,000 yd³ of contaminated soil and battery casings.	Onsite treatment with a chelating agent involved the excavation, separation, and mixing of contaminated materials. Vibrating screens, feed hopper/conveyor, and heavy equipment were used.
Allen Transformer, AR	VI	PCBs	Drums; electrical equipment; debris; soil; and wastewater.	Backhoe, loader, and bulldozer used to excavate contaminated soil and concrete pad; crane used to move transformers.

(continued)

158 Materials-Handling Technologies Used at Hazardous Waste Sites

Site name	EPA Region	Major contaminants	Principal debris/ materials handled	Debris/materials-handling procedure/equipment
Cleve Reber, LA *	VI	Toluene, chlorobenzene, iron, DDT, DDD, benzoic acid	1100 drums and contaminated soil.	Site work involved excavation of drums and soil by using standard heavy equipment.
Crystal City, TX	VI	DDT, DDE, DDD, toxaphene, BHC, 2,4,-D, PCP, parathion, chlordane, arsenic, dieldrin	Excavated and drums	Heavy equipment was used to excavate a 50 ft x 10 ft x 8 ft trench to contain drums crushed by a bulldozer and contaminated soil.
MOTCO, TX *	VI	Vinyl chloride, acids, lead, mercury	Wastewater from waste disposal pit.	The low-pH wastewater from the disposal pit was pumped through a mixture manifold with 50% NaOH.
Pesses Chemical Co., TX	VI	Cadmium, chromium, nickel, lead, copper	1500 drums and debris; 6 inches of soil removed from the site.	Front-end loader was used to stage and load drums onto trucks; water sprays were used for dust control.
Stewco Site, TX	VI	Tetrachloroethane, methyl chloride, naphthalene, cadmium, DDT, arsenic, lead	Pumping wastewater and an oily sludge from 2 unlined lagoons (75 ft x 75 ft x 10 ft and 55 ft x 15 ft x 15 ft).	Wastewater from the 2 ponds was pumped out with a submersible pump with a sock filter to remove oil and suspended solids; an emulsion layer and the sludge layer were solidified with fly ash (mixed by a backhoe) and loaded onto trucks.
Triangle Chemicals, Inc., TX	VI	Acids, corrosives phthalate esters, benzidine, ethylbenzene	1094 crushed drums; contaminated soil	Drums were emptied and contents were bulked into a tanker truck. A front-end loader was used to crush the drums.
B&B Salvage Co., MO	VII	PCBs	300 transformers; 800 tons scrap metal; soil; drums	The site contained a large amount of uncontaminated scrap metal (cars, etc.) that required the use of a trackhoe-mounted grappler to be brought on site to stage and load the metal for delivery to a recycler. Access roads needed to be cleared and graded.

(continued)

Appendix E   159

Site name	EPA Region	Major contaminants	Principal debris/materials handled	Debris/materials-handling procedure/equipment
Broadway Salvage Oil, MO	VII	PCBs	342 tons of contaminated soil and debris.	Excavation was performed with a large excavator; because of isolated "hot spots" at the site, the equipment was moved on clean plywood to minimize cross contamination. Access road was cleared and graded.
Minker-Cul-De-Sac, MO	VII	Dioxin	2420 yd^3 of soil, vegetation, and metal debris.	Soil was excavated with a backhoe and then loaded into a hopper that emptied into 2-yd^3 polypropylene bags, which were picked up by a crane and deposited on a flat-bed truck; structures were decontaminated by using vacuum high-pressure wash and wipe-down cycles.
*Quail Run, MO	VII	Dioxin	Approximately 20,000 yd^3 soil; 600 yd^3 concrete; and approximately 4000 yd^3 debris.	Soil excavation and disposal procedures were similar to those at the Minker sites; concrete was decontaminated by suspending chunks of concrete from a crane while being pressure-washed.
*Solid State Circuits, MO	VII	Trichloroethylene	108,000 gallons wastewater; 1990 tons soil	Excavation of the concrete slab in the basement required the use of a pavement breaker (hoe ram) attachment on the backhoe because of slab thickness (14-18 in.). A work delay resulted in accumulated rainwater that became contaminated and required pumping and disposal.
*B&C Metals, CO	VIII	Radon	Construction debris	Work on the site consisted of the installation of a plenum wall-stack vent system to reduce radon levels.
Eagle Mine, CO	VIII	PCBs	3 transformers; flood water in the mine; 5 capacitors	Diaphragm pumps were used to pump out the water so that the transformers and capacitors could be removed from the mine.
Gene Murren Farm, CO	VIII	Cyanide, zinc, copper, chromium, nickel	21 drums	The drums were loaded by a backhoe onto a trailer for transportation to a hazardous waste disposer.
Mestas Well, CO	VIII	Heavy metals	Excavated soil from 75 ft x 8 ft water line trench.	A backhoe was used to install a water line to a home with a contaminated well.

(continued)

160 Materials-Handling Technologies Used at Hazardous Waste Sites

Site name	EPA Region	Major contaminants	Principal debris/materials handled	Debris/materials-handling procedure/equipment
PDC Spas, CO	VIII	Iso-cyanates, acetone, styrene	62 drums	Drums were loaded by backhoe; 30 empty drums were rinsed, crushed, and taken to a local landfill.
Woodbury Chemical, CO	VIII	Pesticides	Soil grading	A bulldozer and 4.5-yd^3 loader were used to grade the site prior to installing a security fence.
*Burlington Northern Railroad, MT	VIII	B(a)P, pyrene, zinc, nickel, copper, arsenic, naphthalene, fluorene	3280 yd^3 of sludge/soil, 127,000 gal of wastewater.	A pond containing creosote-contaminated sludges was drained by using submersible pumps, and water was stored in a lined pit and storage tank onsite; sludge was excavated by using heavy equipment and stored in a lined pit.
*McColl Site, CA	IX	Organic and sulfur-containing gases	100 yd^3 soil	A portable building was constructed over the site during the excavation by an excavator; a pug mill was used to size and mix soil.
Alaska Battery Enterprises, AK	X	Lead	2900 yd^3 of contaminated soil	Soil was excavated with heavy equipment; a bulldozer was used to break up frozen soil (temperatures ranged from -3 to -10°F).
Ohlson Mountain AC&W, AK	X	PCBs	4 transformers; 3 storage vans; capacitors	Capacitors and transformers were drained and steam-cleaned; PCB liquids were solidified and drummed; flooring in trailers that was soaked with oil was removed and drummed.
Arrcomp Corp, ID	X	PCBs, toluene, methylene chloride, ethylbenzene, xylene, acetone	9700 gallons of contaminated liquids; 137 yd^3 of contaminated soil; 3 tank trucks; 23 storage tanks.	Tanks were cleaned by hooking a compressor to a drum of kerosene and inserting a hose into the tank to break up the sludge; a backhoe and front-end loader were used to excavate the soil and to load plastic-lined dump trucks.
*Pacific Hide & Fur, ID	X	PCBs	582 capacitors; 16 drums; 180 large pieces of debris; 30 yd^3 of contaminated soil	A steel work pad, sump, and curtain were constructed prior to steam-cleaning and pressure washing the large scrap pieces found on the site; a crane was used to stage and load capacitors into overpacks.

(continued)

Site name	EPA Region	Major contaminants	Principal debris/materials handled	Debris/materials-handling procedure/equipment
NW Dust Control Facilities, OR	X	PCBs	33,000 gal of contaminated liquids; 1000 gal of sludge; 2250 pounds of solidified sludge	11 tanks were cut with acetylene torches (CO_2 was pumped into tanks to reduce explosion potential); contaminated liquids were pumped to tanker trucks for disposal; sludges were solidified with sawdust.
American Crossarm & Conduit, WA	X	Creosote PCPs	120,000 gal of oil and contaminated water; 175,000 pounds of soil/debris	Soil was excavated with a backhoe; oil booms and sorbent pads were used to isolate the oil before pumping it to a tanker truck.
Northwest Transfer Salvage Yard, WA *	X	PCBs	1440 yd³ of contaminated soil; 6000 gal of PCB-contaminated liquids; several piles of debris; 4 transformers	Excavation of the soil was accomplished with bulldozers and backhoes; transformers were staged above a holding tank by using a forklift for draining and rinsing; a jackhammer was used to remove a concrete berm; wood inside building was sand-blasted.
Williams & Son Transformer Salvage, WA	X	PCBs	222 yd³ of debris/soil; 1000-gal tank containing sludge; 305 light ballasts	Floor of building was broken up by using a jack-hammer and the pieces disposed of; the 1000-gal tank, with a small amount of sludge, was located on a truck and shipped offsite; soil was excavated with a backhoe and front-end loader.

* Sites for which case studies have been presented in Section 7.

Appendix F

Equipment Costs for Hazardous Waste Work

APPENDIX F

EQUIPMENT COSTS FOR HAZARDOUS WASTE WORK

Equipment	Rent/lease (1990 dollars)			Purchase
	Per day	Per week	Per month	
Excavation/transportation				
Backhoe				
Rubber-tired (1-yd^3)	150-235	585-1060	3320-1870	48,780
Tracked (1.25-yd^3)	495-765	1980-3065	6435-9965	215,000
Front-end loader				
Wheeled (1.5-yd^3)	320-520	1300-2090	4210-6780	66,200
Crawler (4.5-yd^3)	670-950	2670-3795	8685-12,330	221,560
Bulldozer				
65-hp	215-465	850-1920	2770-5730	49,700
140-hp	400-750	1600-2990	5210-9380	134,350
Lowboy				
9-ton	130-185	520-600	1660-1950	NA
20-ton	155-250	615-1000	1990-3245	NA
Crane, hydraulic (1.5-ton)	360-420	1440-1600	4685-5210	107,890
Forklift, 4-wheel-drive (2-ton)	95-165	380-740	1240-2210	37,030
Skid-steer loader	95-185	375-730	1125-2185	12,370
Pump/vacuum unit				
Diaphragm pump (2-in.)	95	380	1160	1300
Trash pump (2-in.)	25-85	70-340	205-1015	1900
Submersible pump (2-in.)	60-100	250-400	745-1250	1810
Vacuum unit (1500-gal)	310-375	1240-1495	4025-4860	NA
Vacuum truck (5000-gal)	380-620	1530-2470	4965-8025	180,000

(continued)

APPENDIX F (continued)

Equipment	Rent/lease (1990 dollars)			Purchase
	Per day	Per week	Per month	
Separation/size reduction				
Drum crusher, explosion-proof (7.5-hp)	NA	1100	NA	10,300
Tire/drum shredder	1800-1900	7,585-12,680	24,700-54,500	40,000-300,000
Vibrating screen	370	1400	4160	NA
Crusher	815	3245	9750	NA
Conveyor (40-ft)	245	970	2910	NA
Miscellaneous				
Hand tools	12	50	140	NA
Non-sparking tools	55	250	745	NA
Pressure washer (2500 psi)	90-355	360-1420	1165-4610	NA
Drum grapper, hydraulic	135-165	540-740	1755-2210	NA
Generator (10-kW)	40-70	160-280	515-850	1600
Air compressor (2-hp)	65-110	260-435	850-1410	1200
Bulking tank (5000-gal)	105	425	1390	3700-7500
Air hammer	20	50	140	630-1180
Chain saw (14-in.)	50	225	NA	190
Cutting torch	70	280	840	250
Barrel cart	40	150	NA	275
Drum punch	60	285	1080	NA

NA - Prices not available.

Appendix G

Equipment Descriptions

This appendix contains information on 36 pieces of equipment that are either used or have potential applications for hazardous waste site work. It includes specifications, features, options, manufacturers, and photographs. Attachments and accessories are also included for some pieces of equipment (e.g., bulldozers and front-end loaders). For most types of heavy equipment a wide variation in the size and working capacity exists. Representative models of these pieces of equipment were chosen based on their suitability for site work. It should be emphasized, however, that other models with different capacities and sizes are available. A representative list of manufacturers was also generated for each piece of equipment in the appendix, but the list is not meant to be all inclusive nor to endorse specific vendors.

INDEX TO APPENDIX G

	Page
Backhoe	168
Backhoe attachments	169
Crane	170
Crawler tractor (dozer, bulldozer)	171
Dredging systems	172
Drum grappler	173
Barrel handler	173
Earth auger	174
Excavator (tired)	175
Excavator (tracked)	176
Flexible bulk containers	177
Hand-held accessories	178
Hazardous waste container bag liners	179
High pressure washer (water laser)	180
Hoe-ram (impact hammer)	181
Horizontal auger (composter)	182
Industrial vacuums	183
Loader [(tired), front-end loader]	184
Loader [(tracked), front-end loader]	185
Loader attachments	186
Mixer (mini-maxcrete)	187
Mobile shear	188
Oil skimmer (rope mop)	189
Portable building	190
Portable holding tank (collapsible)	191
Portable vibrating screen	192
Pugmill (silo mixer)	193
Pumps and pumping systems	194
Radioactive waste containers	195
Rough terrain forklift	196
Shredders	197
Shreddder/compactor system	198
Shredder - mixer	199
Skid steer loader (Bobcat)	200
Skid steer loader attachments	201
Trenching machine	203

Backhoe

Source: PEI Associates, Inc.

Specifications

Overall Length: up to 27 ft
Width: 7 ft, 4 in.
Height: 13 ft, 9 in.
Operating Weight: up to 22,500 lb
Digging Depth: up to 18 ft
Reach: up to 22 ft, 8 in.
Lift Capacity: up to 7300 lb
Power: up to 115 hp (net)

Features

backhoe with hydraulic, self-adjusting, self-equalizing, inboard mounted, wet disk brakes; 4-speed power shift single stage, dual-phase torque converter power shift-reverser; automatic return-to-dig; hydraulic self-leveling loader; loader bucket level indicator; work lights; bucket sizes from 7/8 cu. yd to 1-3/4 cu. yd.

Options

- Turbocharger
- Cab heater
- Extendable dipperstick for backhoe
- 3- or 4-lever, pedal-swing backhoe controls
- Reversible stabilizer pads
- Trackshoes

Manufacturers

John Deere & Co.
J.I. Case Co.
Hitachi
Massey Ferguson

Ditch Witch-The Charles Machine Works, Inc.
Caterpillar, Inc.
Komatsu
JCB

Backhoe Attachments

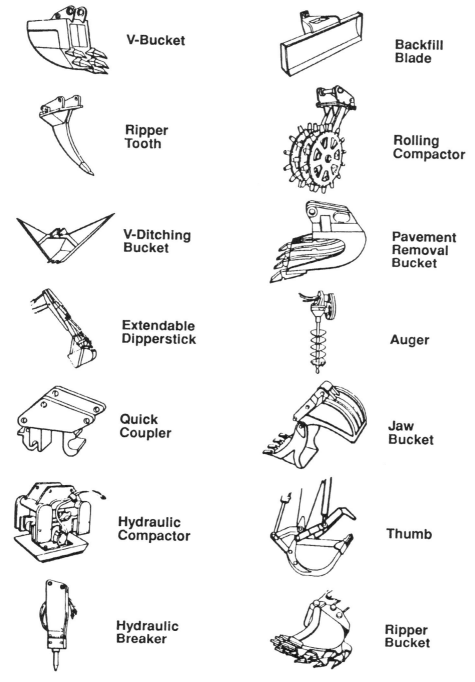

Illustrations courtesy of Deere & Company 1990

Crane

Source: Grove Worldwide Manufacturing 1990

Specifications

Maximum Tip Height: 173 ft
Loaded Boom Angle: 78°
Maximum Load Capacity: 12 to 28 tons
Boom Length: up to 155 ft
Overall Length (stowaway position): up to 41 ft, 9 in.
Overall Width (outriggers extended): up to 24 ft

Features

truck-mounted rough terrain hydraulic crane; four section trapezoidal full-power boom; telescopic swingaway extension (maximum 30° angle); load moment and anti-two block system with audio-visual warning and control level lockout; electronic display of boom angle, length, radius, tip height, relative load moment, maximum permissible load and actual load; ball bearing swing circle with 360° continuous rotation; cold start aid; electronic back-up alarm; engine distress audio-visual warning system; hoist mirrors.

Options

Auxiliary hoist
360° rotating beacon
Engine block heater
Tow winch
Tool kit

Work lights
Cab spotlight
Hookblocks
Spare wheel assembly
Pintle hook front/rear

Manufacturers

Grove Manufacturing Co.
National Crane
Demag Cranes

Abell-Howe
JLG Cranes

Crawler Tractor Dozer

Photo courtesy of Deere & Company 1990

Specifications

Overall Length: up to 18 ft, 6 in.
Height: up to 10 ft, 6 in.
Operating Weight: up to 39,700 lb
Blade Width: up to 138 in.
Power: up to 165 hp (net)

Features

power angle tilt, strait tilt and semi-U blades; oil-cooled steering clutches and brakes; 4-speed direct drive transmission; 4 forward and 4 reverse speeds; lever controls; roll-over protective structures; adjustable blade pitch; chain guides.

Options

- Wide blade
- Wide track
- Halogen work lights
- Counterweights
- Pedal steering

Attachments

Backhoe
Cable plow
Draw bar

Ripper
Sideboom
Winch

Manufacturers

John Deere & Co.
J.I. Case Co.
Hitachi

Caterpillar, Inc.
Komatsu America Corp.

Dredging Systems

Source: Crisafulli Pump Co. 1990

Specifications

Length: 31 ft
Width: 8 ft
Height: 9 ft, 1 in.
Weight (dry): 12,500 lb
Working Depth (std.): 18 ft
Pontoons: (4) 30 in. x 32 in. x 144 in.
Engine: 100, 150, or 177 hp diesel
Pump: 6, 8, or 12 in. hydraulic submersible pump with 8 to 12 in. discharge diameter
Capacity: 1800 to 4500 gpm of water at 40 ft;
1000 to 2500 gpm of sludge at 40 ft

Features

modular dredging system with hydraulic control; travels from 0 to 100 ft per minute; propelled by two-, four-, or windlass-winch system; can control percentage of solids.

Options

- Removable power unit
- Cab
- Canopy
- Slurry gate pump
- Cutterheads
- Jetting ring
- Air-conditioning
- Heated cab

Manufacturers

Crisafulli Pump Co.
VMI, Inc.
H&H Pump Co.

Aquatics Unlimited
Liquid Waste Technology, Inc.

Appendix G   173

Drum Grappler

Barrel Handler

Source: PEI Associates, Inc.

Source: LaBounty Manufacturing, Inc. 1990

Specifications

Height: 52 in.
Weight: 1,700 lb
Width: 30 in.
Open: 40 in.
Close: 22 in.

Features

specially designed barrel handlers for moving barrels, including those of hazardous waste; 360° rotating turntable mechanism; 3/4 in. non-sparking neoprene lining; made with high-alloy, high-tensile, abrasion-resistant steel; heat-treated alloy pivot bearings; fits most excavators and backhoes.

Manufacturers

LaBounty Manufacturing, Inc.
Gensco Equipment Co. Ltd.
Mack Manufacturing, Inc.
Downs Crane & Hoist Co., Inc.

Earth Auger

Source: Ditch Witch 1990

Specifications

Length: 108 in.
Width: 30 in.
Height: 43 in.
Auger Speed: 0 (min.) to 90 (max.) rpm
Case Bore Diameter: 4 to 20 in.
Maximum Thrust: 14,000 to 48,000 lb
Operating Weight: 650 to 1450 lb

Features

hydraulically-powered earth augers allow variable boring speeds; will work in boring pits containing water; can bore through most soil conditions; power supplied by power sources located outside the boring pit for reduced engine heat and noise.

Also Available

Pneumatic pipe pushers

Options

Variety of cutterheads & augers for various types of soil

Manufacturers

Ditch Witch - The Charles Machine Works
Allied Steel & Tractor Products, Inc.

Excavator (Tired)

Photo courtesy of Deere & Company 1990

Specifications

Overall Length: up to 28 ft, 3 in. (std. arm)
Width: 8 ft, 2 in.
Height: up to 10 ft, 2 in.
Operating Weight: up to 36,575 lb
Digging Depth: up to 20 ft, 8 in.
Reach: up to 31 ft, 4 in.
Lift Capacity at 20 ft: up to 10,150 lb
Power: 95 hp (net)

Features

excavator with turbocharged diesel engine; high efficiency, variable flow hydraulic system; two-lever, low effort pilot controls; 2 speed choices (max speed 21.4 mph).

Options

- Short or long arm
- Bucket sizes from 1/2 cu. yd to 3/4 cu. yd
- Dual or single tires
- Rear blade or stabilizers
- Remote control operation
  - up to one mile without visual contact
  - 8 hours of operation on batteries in operator control units
  - remote operation via radio frequency or hardwire cable controls
  - high resolution video & audio feedback
  - proportional control of dig functions
  - discrete on/off control of engine, dozer blade, travel speed

Manufacturers

John Deere & Co.
Gradall Co.
Caterpillar, Inc.

Vermeer Mfg. Corp.
Saf-T-Boom Corp.
Komatsu

Excavator (Tracked)

Photo courtesy of Deere & Company 1990

Specifications

Overall Length: up to 38 ft, 7 in. (std. arm)
Width: up to 11 ft, 11 in.
Height: up to 10 ft, 8 in.
Operating Weight: 97,350 lb
Digging Depth: up to 26 ft, 10 in.
Cutting Height: up to 36 ft, 3 in.
Dumping Height: up to 25 ft, 6 in.
Power: 265 hp (net)

Features

excavator with turbocharged and aftercooled diesel engine; two-lever, low effort hydraulically controlled boom, arm, bucket, and 360° swing; straight propelling; track-type undercarriage with sealed track chain; hydraulic track adjustment; 3 digging mode selections; work lights.

Options

- Short or long arm
- Extra wide track shoes
- No-arm attachment option
- No-boom attachment option
- No-bucket attachment option
- Bucket sizes from 1/2 cu. yd to 2-3/4 cu. yd

Manufacturers

John Deere & Co.
Gradall Co.
Caterpillar, Inc.

Vermeer Mfg. Corp.
Saf-T-Boom Corp.
Komatsu

Appendix G   177

Flexible Bulk Containers
Collapsible Containers

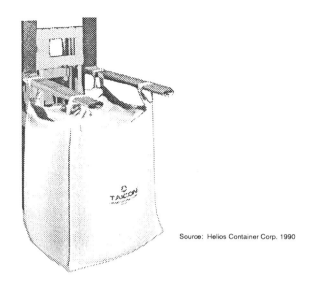

Source: Helios Container Corp. 1990

Specifications

Volume: 15 to 60 cu. ft
Sizes (L x W x H): 35 in. x 35 in. x 18 in. to 35 in. x 35 in. x 69 in.
Std. Tops: duffel (full open) and charge spout (14 in.)
Std. Bottoms: closed (1-way use), discharge spout (16 in.) and full open

Features

collapsible, self-emptying container with top loading spout and optional bottom dispenser; can feed directly into processing equipment; can use forklift to load/unload containers; can be used with conventional material handling systems; can be loaded onto truck, rail car, and barge; can be used for transporting or storing chemicals, fertilizers, minerals.

Options

- Custom-sized rectangular bags
- Custom-sized cylindrical bags (44 in.)
- Silkscreen company name on bag
- Document pouch, warning labels sewn on

Manufacturers

Helios Container Systems, Inc.
Bulk-Lift International, Inc.

UF Strainrite
Walpole, Inc.

Hand-Held Accessories

Jack Hammer
Height: 23 to 27 in.
Blows Per Minute: 800 to 2500
Working Pressure: 100 to 1950 psi
Maximum Back Pressure: 145 psi

Two-Man Auger
Flow: 5.3 gpm
RPM: 90 (maximum), forward or reverse
Torque: 270 ft-lb
Pressure: 1500 to 2200 ft-lb
Weight: 70 lb

Cut-Off Saw
Length: 19 in.
Wheel Diameter: 14 in.
Weight: 18.5 lb
RPM: 4700 to 5300
Arbor Size: 1 in.

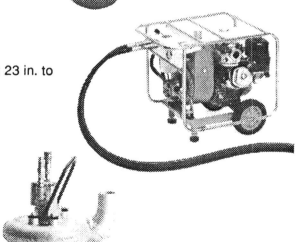

Power Pack
Engine: 8 to 11 hp (gasoline)
Dimensions (L x W x H): 26 in. x 19 in. x 23 in. to 34 in. x 22 in. x 26 in.
Fuel Capacity: 1.2 to 1.8 gal
Weight: 121 to 187 lb

Submersible Pump
Weight: 21 to 62 lb
Height: 10.5 to 17.3 in.
Width: 8.5 to 18 in.
Discharge Size: 2 to 3 in.
Output (max): 225 to 670 gpm
Solid Size: 0.375 to 1.75 in.
Power Requirement: 5 to 7 hp

Source: Allied Steel and Tractor Products, Inc. 1990

Manufacturers
Allied Steel & Tractor Products, Inc.
Ingersoll-Rand Co.

Hazardous Waste Container Bag Liners

Source: Packaging Research and Design Corp. 1990

Specifications

Thickness: 4, 6, 8, or 10 mil
Bottom Width: 92 or 96 in.
Height: up to 120 in.

Features

constructed of plastic made from a special blend of resins; highly resistant to tears, punctures and chemicals; designed to fit inside roll-offs and dump trailers; snug corner fit; most can be installed by one person in 5 minutes; extra-tall liners (96, 100 or 120 inches) can be folded to center and sealed for transporting sludges, dust, odorous materials, asbestos; form fits to container size.

Options

- 3 grades of plastic: high performance (heavy-duty) to industry average

Manufacturers

Packaging Research and Design Corp.
Specialty Plastic Fabricators, Inc.
Flexilon Package Corp.

High Pressure Washer
Water Laser

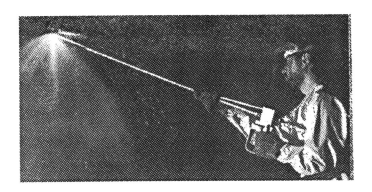

Source: Allied Steel & Tractor Products, Inc. 1990

Specifications

High Pressure/Low Flow Pump Head
Pressure: up to 40,000 psi
Flow: 1 to 46 gpm

Low Pressure/High Flow Pump Head
Pressure: up to 2950 psi
Flow: up to 128 gpm

Features

pneumatically operated hand-held tool which can be used as a cutting tool or for equipment decontamination; cuts concrete up to 6 in. in thickness; cold cutting of ferrous and non-ferrous metals.

Manufacturers

Allied Steel & Tractor Products, Inc.
Ameriquest Co.
Valley Hydro/Laser Technologies, Inc.
Laser Applications, Inc.
Ingersoll-Rand Co.

Hoe-Ram Impact Hammer

Source: Allied Steel & Tractor Products, Inc. 1990

Specifications

Overall Length: 75 in.
Working Length: 17 in.
Chisel Diameter: 3.5 in.
Weight with Std. Bracket: 1200 lb
Impact Energy Class: 1200 ft-lb
Frequency: 500 blows/min.
Operating Pressure: 100 lb/sq. in.

Features

boom-mounted impact hammer; standard cross-cut chisel demolition tool; automatic shut-off at any angle; easily removable boom pins for hammer-to-bucket exchange; ideal for pavement breaking, rock and boulder removal, building demolition, foundation removal, cleaning slag pits.

Also Available

truck-mounted, self-propelled hydraulic impact hammer; hydraulic-tilt chassis to work on uneven surfaces; choice of breaking, cutting, or tamping tool.

Options

- Tool variations
- Hydraulic or pneumatic

Manufacturers

Allied Steel & Tractor Products, Inc.
Ingersoll-Rand Co.
Bomag Corp.
Broderson Mfg. Corp.

Horizontal Auger Sludge Aerator Composter

Source: Brown Bear Corp. 1990

Specifications

Length: 39 in.
Height: 51 in.
Overall Width: 81.5 in.
Swath Width: 72 in.
Tilt: 10° either direction
Discharge: left hand
Weight: 1300 lb
Handling Capability: 100 to 4000 tons per hour

Features

self-contained auger/aerator attachment for loaders not exceeding 8000 lb; reversible, replaceable blade; 24 in. diameter auger or 24 in. paddle aerator; adjustable dirt shield.

Options

- 24 in. dia. auger with 4 position, quick adjusting wear plates
- 24 in. dia. compound helical paddle auger
- Quick attach mounts for most vehicles

Manufacturers

Brown Bear Corp.
SWR Corp.

Industrial Vacuums

Source: Super Products Corp

Specifications

Length: 161 to 216 in.
Height: 124 to 132 in.
Width: 96 in.
Weight: 7,160 to 11,500 lb
Air flow: 1,350 to 4,500 cfm
Maximum rated vacuum: 200 in. H_2O
Air flow at maximum vacuum: 1050 to 3750 cfm

Features

positive displacement vacuum pump; exhaust silencer; diesel engine; hinged top access doors for baghouse maintenance; wet or dry vacuum; ground level access door for maintenance of first stage baghouse; permanently mounted, skid mounted, or wagon-mounted.

Options

Sound Suppression Package
Engine Safety System
Level Detector
Work Lights
Hazardous Waste Containers

Beacon Lights
Diesel-Powered Air Compressor
Various Diesel Engines
Tapered Baghouse
Removable Collection Bin

Accessories

HEPA Filters
Fluidized Nozzles
Floor/Wall Nozzles
Crevice Nozzles

Suction Hose
Hose Couplers
Permanent Ducting Systems

Manufacturers

Super Products Corp.
DeMarco Max Vac Corp.
Vacuum Engineering Corp.

Ultravac Products
Edwards High Vacuum, Inc.

Loader (Tired)
Front-End Loader

Photo courtesy of Deere & Company 1990

Specifications

Overall Length: up to 26 ft, 6 in.
Height: up to 14 ft, 5 in.
Operating Weight: up to 45,640 lb
Tipping Load, straight: up to 34,500 lb
Tipping Load, 40° turn: up to 29,140 lb
Dump Clearance, 45° dump: up to 117.12 in.
Reach, 45° dump/full height: up to 54.17 in.
Power: up to 216 hp (net)
Bucket Capacities: up to 5.0 cu. yd

Features

turbocharged diesel engine; four wheel drive; power steering and power brakes; single-lever control; roll-over protective structures; maneuverable in tight spaces; power-shift transmission; forward and reverse speeds.

Options

- Automatic return-to-dig and boom height control
- Cold-weather starting aid
- Work lights
- Heater and A/C
- Rear counterweights

Attachments

Bucket teeth
Bolt-on cutting edge
Snow plows

Forks
Special buckets

Manufacturers

John Deere & Co.
J.I. Case Co.
Komatsu Corp.

Caterpillar, Inc.
Terramite Corp.
JCB

Crawler-Loader (Tracked) Front-End Loader

Photo courtesy of Deere & Company 1990

Specifications

Overall Length: up to 18 ft, 6 in.
Height: up to 10 ft, 3.7 in.
Operating Weight: up to 37,450 lb
Tipping Load: up to 25,340 lb
Dump Clearance: up to 112 in.
Bucket Size: up to 2.25 cu. yd
Power: up to 140 hp (net)

Features

automatic hydrostatic drive; single lever travel speed and direction control; turbocharged diesel engine; pedal controls; roll-over protective structures; counter-rotating tracks; single-lever loader control with float position and return-to-dig; boom safety lock pin; push button starting.

Options

- Cab heater and A/C
- Bolt-on bucket teeth
- Front pull hook
- Halogen work lights
- Cold-weather starting aid

Attachments

Winch
Ripper
Cable plow

Multipurpose bucket
Drawbar

Manufacturers

John Deere & Co.
J.I. Case Co.
Komatsu Corp.

Caterpillar, Inc.
Terramite Corp.
JCB

Loader Attachments

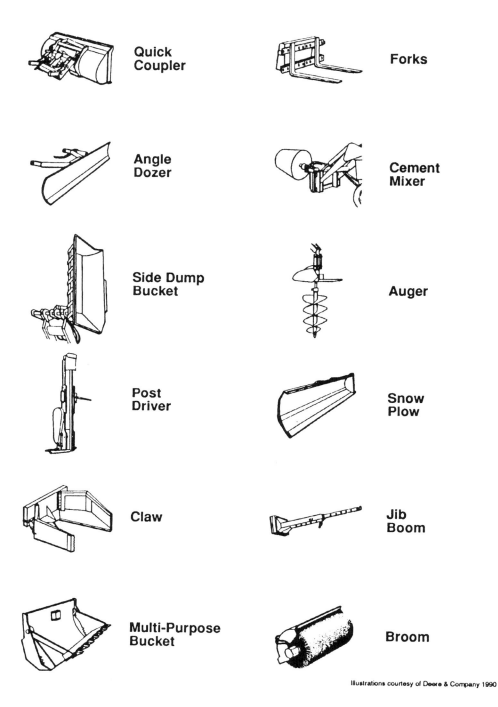

Illustrations courtesy of Deere & Company 1990

Mixer
Mini-Maxcrete

Source: Maxon Industries, Inc. 1990

Specifications

Overall Length: 13 ft, 2-7/8 in.
Overall Width: 6 ft, 9 in.
Overall Height: 6 ft, 8-1/2 in.
Mixing Speed: 2-10 rpm (rotates in two directions)
Capacity: 410 gal
Power Unit: 18 hp Briggs & Stratton gasoline engine
Hydraulics: 2500 psi rated circuit with high pressure gear-type pump

Features

mixer-agitator with double reduction chain drive with high pressure orbital hydraulic motor driving heavy-duty shaft with 8 urethane paddles; tilt-away grid top for loading, visual mixing inspection; 3-position mixer control (charge/stop/discharge); gate has double-acting cylinder with inching control for metered discharge.

Options

Barrel Lift: 1000 lb lifting capacity with heavy-duty lifting cylinder; also tilts and swings
Power Unit: 10 hp electric motor available

Manufacturers

Maxon Industries
Philadelphia Mixers Corp.

VFL Technology Corp.
Davis Pugmill Inc.

Mobile Shear

Source: LaBounty Manufacturing, Inc. 1990

Specifications

Cutting Depths: 10 to 76 in.

Features

mobile shear constructed of high tensile, high alloy, abrasion-resistant steel; processes steel beams, reinforced steel pipe, rail cars, tree stumps and tires; 360° continuous rotation; high performance cylinders; angle actuators; range in sizes to fit skid-steer loaders to excavators.

Manufacturers

LaBounty Manufacturing, Inc.

Appendix G   189

Oil Skimmers
Rope Mop

Source: Oil Mop, Inc. 1990

Specifications

Length: up to 10 ft
Width: up to 7.5 ft
Height: up to 7 ft
Weight (dry): up to 8000 lb
Mop Speed: 0-150 ft/min
Mop Length (max): 2000 ft
Power: 37.5 hp at 1800 rpm
Recovery Capacity: up to 200 Bbl/hr+

Features

skid-mounted wringer units with oil collection pan; guide and training rollers; squeezer rollers; hydraulically powered pumps; centrifugal trash pump used to transfer oil from collection pan; mop speed & direction controlled by lever connected to pump; mop made with oleophilic, hydrophobic fibers.

Options

- Range of sizes from barrel-mounted wringers to large skid-mounted units

Manufacturers

Oil Mop, Inc.
Containment Systems
Abanaki Corp.

Oil Skimmers, Inc.
Hudson Industries

Portable Building

Source: PEI Associates, Inc.

Specifications

Spans: 30-115 ft
Length: any, in modules of 10 ft
Height: variable

Features

prefabricated modular buildings for rapid erection and deployment; hot-dip galvanized steel tube arch frames covered with PVC-coated polyester fabric which is self-extinguishing & fire retardant; may also be constructed of coated aluminum; custom-manufactured sizes available; can easily be extended or sub-divided; no internal columns; double skin options for increased insulation; can be lifted in erected form by crane; designed to withstand foul weather (Arctic and tropical conditions); climate-controlled.

Options

- Choice of color combinations

Manufacturers

Rubb, Inc.
Canvas Specialty

Anchor Industries, Inc.
Spandome Corp.

Portable Holding Tank
Collapsible Storage Tank
Fabric Tanks, Bladder Tanks

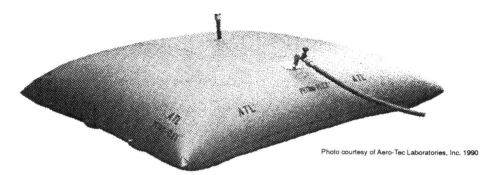

Photo courtesy of Aero-Tec Laboratories, Inc. 1990

Specifications

Sizes: 16 different capacities from 100 gal to 100,000 gal
Access: 2 in. fill/discharge fitting & access plate
Ventilation: vent pipe, spark arrestor, pressure relief
Fittings: steel and aluminum unless otherwise specified

Features

collapsible, rubberized containers which lay flat when empty and assume pillow configuration when full of liquid; containers collapse around liquid, eliminating vapor build-up and evaporative emissions; limits liquid losses; keeps out dust and other contaminants; compacts to less than 5% of expanded shape; resists outdoor exposure to sunlight (UV light), high and low temperatures, abrasion, and mildew; can hold industrial chemicals (alcohol, acids, alkalis, aqueous salt solutions); not resistant to chlorinated solvents, ketones, aromatics.

Options

- Intermediate sizes on custom order
- Special shapes, fittings or color on special order

Accessories

Flanges	Pumps	Repair Kits	Couplings
Fittings	Gate Valves	Ground Cloths	Nozzles
Manholes	Check Valves	Hoses	Filters

Manufacturers

Aero-Tec Laboratories, Inc.
Kepner Plastics Fabricators, Inc.

Vinyl Tech, Inc.
Terra Tank Corp.

Portable Vibrating Screen

Source: Powerscreen of America 1990

Specifications

Length: up to 28 ft
Height: up to 11 ft, 6 in.
Width: 7 ft, 11 in.
Weight: up to 24,200 lb
Screen Size: up to 12 ft x 7 ft
Hopper Opening: up to 14 ft, 9 in.
Conveyor Belt Width: 5 ft
Conveyor Belt Speed: 270 ft/min.
Max. Production: 650 tph

Features

screen powered by a 47 hp air-cooled diesel hydraulic unit; twin road wheels, heavy-duty tow bar and hydraulic jacking leg; screen mesh sizes range from 1/4 in. to 4 in.; screening angle adjusts from 12° to 25°.

Manufacturers

Powerscreen of America, Inc.
Read Corp.
Kason Corp.
FMC Corp.
Simplicity Engineering

Appendix G 193

Pug Mill
Silo Mixer

Source: PEI Associates, Inc.

Specifications

Load-Out Hopper Capacity: 20 cu. yd
Maximum Output: 800 tph
Motor: 200 hp with automatic transformer starter
Pug Mill: KA-85 Barber-Greene style twin-shaft
Hydraulics: 400 gpm water-injection system
Conveyor: 4 ft (width) x 70 ft (length)
Clearance Below Load-Out Hopper: 15 ft

Features

computerized pug mill with paddle-type mixing tips mounted on pug mill shafts; designed for continuous operation (no batching); automatically adjusts rate of input after weighing material by changing belt speed and/or hopper opening resulting in a consistent product; shredders in each input hopper; computer printout of daily activities; shaft assemblies are mounted in heavy-duty bearings with shaft seals mounted on each end to prevent leaking; stores in memory up to 50 mixing designs with 6 additives.

Manufacturers

Conti Construction Co.
Davis Pugmill, Inc.
Eastern Cleveland Mixers
Marion Mixers, Inc.

Pumps and Pumping Systems

Source: Crisafulli Pump Co. 1990

Submersible Pump

Discharge Sizes: 2 to 16 in.
Flow: 100 to 9000 gpm
Motor: electric, diesel, gas/hydraulically or electrically driven pumps

Vertical Pump

Discharge Sizes: up to 24 in.
Flow: up to 20,000 gpm
Will pump high weight fluids.

Trailer Pump

Discharge Sizes: 4 to 24 in.
Flow: 500 to 20,000 gpm
Motor: diesel or electric
Will handle thick sludges and mud.

Slurry Pump

Discharge Sizes: 6 to 16 in.
Flow: up to 3800 gpm
Mixing to 11,000 gpm
Will handle highly viscous fluids.

- All pumps can be custom-made.

Manufacturers

Crisafulli Pump Co.
Wilden Pump & Engineering Corp.
H&H Pump Co.
Edson International

Radioactive Waste Containers

High Integrity Containers: Type A Quantities

Source: Scientific Ecology Group, Inc.

Specifications

Internal Volume: 6.5 to 173 cu. ft

Features

storage and transport container constructed of high density cross-linked polyethylene; attachable grapple beams for remote grappling purposes; may be equipped with steel liners and/or underdrains.

Shipping Casks: Type B Quantities

Source: Scientific Ecology Group, Inc.

Specifications

Internal Dimensions (dia. x height): 54 in. x 62.13 in. to 76.75 in. x 80.25 in.
Lead Shielding Equivalence: 1.81 to 4.58 in.
Approximate Maximum Rad Level: 3 to 800 rem/h
Maximum Payload: 20,000 lb

Features

re-inforced shielded casks for shipping radioactive waste; some equipped with impact rings constructed of high density polyurethane foam to withstand 30 ft drop; thermal insulation to withstand thermal radiation requirements.

Manufacturers

Scientific Ecology Group, Inc.
Electropanel

Rough Terrain Forklift

Source: Massey Ferguson Industrial 1990

Specifications

Length (less forks): up to 165 in.
Width: 60 to 84 in.
Height to Overhead Guard: 98 in.
Turning Radius: 115 to 168 in.
Lift Capacity: up to 100,000 lb
Lift Height: up to 30 ft
Mast Tilt: 15° forward/12° backward

Features

48 in. forks; front work lights; brake lights; cold start assist; foot and hand throttle; differential lock; roll-over protection system; side shifter mast; instrument and warning lights; horn.

Also Available

variable reach forklift with 37 ft, 6 in. reach; horizontal & vertical fork movement; telescopic 3-stage welded box section boom.

Options

- 4-wheel drive
- Enclosed cab with heater, defroster, wipers & mirror
- Rear work lights
- Fork lengths: 42 to 96 in.
- Mast height: 10 to 30 ft

Manufacturers

Massey Ferguson Industrial Machinery, Inc.
Trak International
Caterpillar, Inc.

Clark
Komatsu

Shredders

Source: MAC/Saturn Corp. 1990

Specifications

Length: up to 150 in.
Width: up to 72 in.
Infeed Opening (L x W): 22 in. x 22 in. to 120 in. x 58 in.
Power: up to 500 hp
Shaft Torque Range: up to 80,000 ft-lb
Cutter Thickness: 5/8 in. to 4 in.
Cutter Diameter: 5-1/2 in. to 30 in.

Features

hydraulic-driven rotary shear-type shredder; low noise; low dust; automatic reverse, non-jamming capability; low speed and high torque; replaceable cutters, spacers, cleaning fingers, keys; cutting accomplished by drawing material past interfaces of 2 counter-rotating blades in close tolerance; can handle hazardous/nuclear waste, tires and rubber, ferrous and non-ferrous materials, solid waste and batteries.

Options

- Stationary or portable configuration
- Diesel or electric motors

Manufacturers

MAC/Saturn Corp.
Shredding Systems, Inc.
Hi-Torque Shredder Co.

Jersey Stainless, Inc.
Eidal International Sales Corp.

Shredder/Compactor System

Source: MAC/Saturn Corp. 1990

Features

shear-type shredder combined with a high density compactor; automatically shreds and compacts plastics, metals (including drums and conduit), wood, cardboard, rubber, cloth, paper, etc; bales are 15 in. x 30 in. x variable length; 3 bales automatically loaded into a 45 cu. ft container.

Manufacturers

MAC/Saturn Corp.
S&G Enterprises, Inc.
International Baler Corp.
Compaction Technologies, Inc.
Consolidated Balers & Compactors

Appendix G   199

Shredder-Mixer Earth Materials Processor Composter

Source: Powerscreen of America 1990

Specifications
Width: 8 ft
Height: 13 ft, 6 in.
Maximum Rated Capacity: 200 cu. yd/h
Hopper Opening: 7 ft x 12 ft
Hopper Capacity: 5 cu. yd
Discharge Height: 12 ft, 8 in.

Features
conveyor and stone grate; hydraulic brakes; highway tires; shredding belt; lump breakers; trash-away conveyor; variable sweep and deflector.

Options
- Gas or diesel engine
- Programmable controller (adjusts feed rate for maximum load, reverses shredding belt at selected intervals to clean debris, operates shaker gate, safety switches)
- Shaker gate
- Hopper platform

Manufacturers
Powerscreen of America/Royer Industries
Compost Systems Co.

Skid-Steer Loader

Source: Melroe Company 1990

Specifications

Length (less bucket): 73 to 118 in.
Width (less bucket): 36 to 90 in.
Height: 72 to 96 in.
Height to Bucket Pin: 92 to 130 in.
Rated Operating Capacity: 600 to 4000 lb
Tipping Load: 1217 to 8100 lb

Features

small loader with multi-job attachment versatility; hydraulic bucket positioning; work lights; operator cab; seat bar; seat belt; instrument panel; tip-up operator cab; gasoline or diesel engine; 22 gal fuel tank; spark arrestor muffler; hydrostatic transmission.

Also Available

a multipurpose tool carrier (MTC) with four-wheel hydrostatic drive which provides a tighter turning radius.

Options

- Enclosed cab
- Flotation tires
- Lift arm stops
- Cab heater
- Suspension seat
- Propane fuel source

Manufacturers

Bobcat-Melroe Co.
John Deere and Co.
J.I. Case Co.
New Holland, Inc.
Gehl

Appendix G   201

Skid Steer Loader Attachments

Backhoe
- High production digging
- Full line of models for a variety of loader sizes

Combination Bucket
- Ideal for dozing, grappling, leveling, digging, loading and dumping
- Heavy-duty cutting edges and cylinders

Industrial Fork Grapple
- Handle scrap and assorted metal, including irregular shapes

Hydraulic Breaker
- For concrete demolition
- Up to 750 ft-lb impact energy class
- Backhoe or loader mount

Landscape Rake
- Grade, level, and scarify in close quarters
- Break up lumpy soil or pick up rocks and dump them where needed

LaBounty Shear
- Cut through rebar, scrap metal, concrete, pipe
- 360° rotation

Angle Blade
- Snow removal
- Dozing and backfilling dirt

Pallet Fork
- Handle baled or pallet materials
- Move bulky or bagged materials

Longwood Grapple
- High performance log handler

Grapple Bucket
- Move heavy and odd-shaped objects with ease
- Handle scrap, waste, or pipe

Source: Melroe Company 1990

Skid Steer Loader Attachments
(continued)

Grader
- 7 ft, six-way hydraulically controlled moldboard
- Used for landscaping and asphalt/concrete work

Earth Auger
- Hydraulic auger with high-torque, low-speed power
- Dig holes 6 to 36 in. deep

Scarifier
- Adjustable depth skids for preset digging depth
- Rips asphalt for removal

Sweeper
- Clean up, spread sand, or scrape mud
- Sweep and collect debris with one attachment

Tire Tracks
- Float over soft, sandy, or muddy ground
- Increase traction on slippery surfaces

Hydraulic Trencher
- Trench with side-to-side maneuverability
- 4 to 12 in. widths and up to 48 in. deep

Utility Fork/Grappler
- Useful for handling bundled material, loose straw

Vertical Mast
- For stacking, lifting, or loading material
- 2 models: 10 or 12 ft lifting height

Vibratory Roller
- Smooth or padded 48 in. wide drum
- 4156 lb of force at 2153 vibrations per minute

York Rake
- Rip up soil with scarifier teeth, blade down high spots & fill holes, rake surface to remove large rocks and debris

Source: Melroe Company 1990

Trenching Machine

Source: Ditch Witch 1990

Specifications
Attachment Length: 7 ft, 11 in.
Overall Length: 15 ft, 11-3/4 in. to 19 ft, 2-1/4 in.
Operating Height: 95 to 98-1/2 in.
Width (with backfill blade): 64 to 77 in.
Maximum Plow-In Depth: 18 to 30 in.
Maximum Trenching Depth: 48 to 62 in.
Maximum Trenching Width: 12 in.

Features
combination of trencher and vibratory plow in one attachment; fits 30 to 65 hp class Modularmatic® equipment; feed-in or pull-out vibratory plow blades; can be used in combination with other attachments.

Also Available
- Front-end mounted backhoe with a maximum digging depth of 111 in.
- Trenching machine which digs trench and installs flexible drain tubing and filter medium in one continuous operation.

Options
- Double auger system
- Heavy duty digging chain
- Offset trencher
- Remote handling apparatus

Other Attachments
Double-pivot trencher
Trencher
Vibratory plow
Reel carrier

Earth saw
Hydraulic boring units
Roto-Witch

Manufacturers
Ditch Witch - The Charles Machine Works
Gradall Co.
Bobcat - Melroe Co.

Vermeer Mfg., Inc.
Industrial Builders, Inc.

Glossary

aggregate: Inert material (mineral), such as sand, gravel, shells, slag, or broken stone, used for mixing to form concrete.

air hammer: A machine hammer driven by an air compressor (e.g., a jackhammer).

amortization: Repayment in equal installments of an accumulated debt, principal and interest. Allows for gradual recovery of the cost of expensive equipment.

angledozer: A power-operated machine fitted with a blade that is adjustable in height and angle for pushing, sidecasting, and spreading loose excavated material as for open cast pits; clearing land; or leveling runways.

auger: A wood- or soil-boring tool that consists of a shank with spiral channels ending in two spurs, a central tapered feed screw, and a pair of cutting lips.

backhoe: An excavating machine with a hoe-type or pull-type shovel. May be rubber-tired or tracked.

batch: A specific quantity of material prepared, loaded, or excavated per unit operation.

bearing capacity: Maximum ability of a material to support an imposed load before failure (i.e., collapse).

berm: A horizontal ledge cut between the foot and top of an embankment to stabilize the slope by intercepting sliding earth.

blade: The broad flat or concave (U-shaped) part of an excavating machine used to dig or push rock/earth (but not to carry it).

bobcat: See skid-steer loader.

boom: Temporary floating barrier used to contain oil and catch floating objects.

bucket: The typically rounded part of an excavating machine used for digging, lifting, and carrying material.

bulking tanks: Plastic or metal receptacles used for storing and/or mixing material (usually liquid) at hazardous waste sites.

bulldozer: Crawler tractor with a hydraulic or cable-controlled front-mounted blade.

capping: The process of covering buried waste materials with a cover material (usually clay).

CAT: Trademark name for the Caterpillar Tractor Company.

cherry picker: A small truck or tractor-mounted derrick or hoist.

clamshell: A two-jawed bucket mounted on flat-bottomed barges or crawler tractors. Loads by its own weight when lifted by a line.

classification: Arranging or sorting waste materials into uniform categories or classes (usually by size, weight, color, shape, etc.). (See separation, screening.)

compaction: Reduction in volume of material--generally of fill or soil--by rolling or tamping.

conveyor: A mechanical device used to haul materials by cable, belts, or chains.

crane: A machine used for lifting, transporting, and lowering loads. Also used for the handling of buckets for excavation and dredging.

crawler: A set of roller-chain tracks used to support and propel an excavation machine.

crusher: A machine used to break up material into smaller sized pieces by a pounding action with hammers, beaters, etc.

cubic yard loose measurement (cylm): Unit of excavation in machine, in stockpile, or after compaction (Church 1981).

cubic yard struck measurement (cysm): Unit of capacity of bucket body, bowl, or dipper of a machine. Sometimes called water-level measurement (Church 1981).

cutterhead: A set of revolving blades used for cutting--set at the beginning of the suction line of a hydraulic dredge.

debris: Unused, unwanted, or discarded solids or liquids that require staging, loading, transporting, pretreating, treatment, or disposal at a hazardous waste site.

dewatering: The removal of water by filtration, evaporation, centrifugation, thickening, pressing, pumping, or draining.

dragline: An excavation machine with a revolving shovel that carries a bucket attached by cables and digs by pulling (scraping) the bucket toward the machine.

dredge: A machine used for the excavation of soil/sediment from the bottom or the banks of a body of water.

excavation: Removal of the ground surface by cutting, digging, or scooping.

excavator: A machine for digging and removing earth with a hoe attachment; generally larger than backhoes.

ferrous: Metals that are primarily composed of iron (steel or tin cans, automobiles, refrigerators, etc.).

flash point: The lowest temperature at which a material will volatilize to yield sufficient vapor to form a flammable gaseous mixture with air.

fly ash: Small particles of ash and soot generated as a result of burning coal, oil, or waste materials.

forklift: A self-propelled machine (usually gas or diesel) used to lift and transport objects.

front-end loader: A loader with a front-mounted hydraulically operated bucket; used for excavation and hauling (payloader).

grappler: Hydraulic drum grapple attachment that uses heavy tongs and a wrist-action motion and 360-degree rotation along the plane of attachment to a backhoe.

grizzly: A set of parallel bars frame-mounted on an angle to promote the flow and separation of excavated material.

groundwater: Water occurring in the zone of saturation in an aquifer or soil.

hammermill: A mechanical device used to break up waste material into smaller pieces by use of a system of heavy rotating hammers.

hazardous waste: "...A solid waste or combination of solid wastes, which because of its quantity, concentration, or physical, chemical, or infectious characteristics may -

(A) cause or significantly contribute to an increase in serious irreversible, or incapacitating reversible illness;

or

(B) pose a substantial present or potential hazard to human health or the environment when improperly treated, stored, transported, or disposed of, or otherwise managed." [RCRA 1004(5)].

hoe: Any of various instruments used for tilling, mixing, or raking.

hoe ram: Hydraulic or pneumatic impact hammer used for breaking up rock, concrete, asphalt, etc., which can be mounted on a backhoe or skid-steer loader.

hopper: A funnel-shaped receptacle with an opening at the top for loading and a discharge opening at the bottom for bulk-delivering material.

hydraulic: Operated or affected by the action of water or other fluid of low viscosity.

jaw bucket: A hoe bucket backhoe attachment with a rear hinge for easier unloading.

jaw crusher: Primary crusher with a fixed and an oscillating jaw spaced at the top and closely spaced at the bottom.

jib boom: An extension that is hinged to the upper end of a crane boom.

leachate: A liquid solution containing decomposed waste, bacteria, and potentially hazardous contaminants that drains from landfills or contaminated sites.

nonferrous: Metals containing no iron, such as aluminum cans, copper wire, brass, etc.

overburden: The rock, earth, waste materials, or other matter found directly above the material to be excavated (e.g., earth over a buried tank).

overpack: Oversized drum (usually 85-gallon) used for storage of damaged and/or leaking 55-gallon drums.

pneumatic: Pertaining to or operated by air or other gas.

pressure washer: Device used to direct and spray water at high pressure for decontamination of heavy equipment.

pug mill: A machine for mixing and tempering a plastic material by the action of blades revolving in a drum or trough.

resource recovery: The extraction and utilization of valuable material from the waste stream (e.g., glass, aluminum, paper).

ripper bucket: A toothed hoe bucket backhoe attachment for digging in frozen ground, shale, and rock.

riprap: Heavy rocks placed along an embankment of a body of water to prevent further soil erosion.

rolloff box: dumpster box that can be detached from truck chassis and left on site.

scalping: Process of removing residual or weathered rock at a cut prior to excavation.

scarifier: An implement or machine with downward projecting tines for breaking down a road surface or earth 2 feet or less.

screening: Separating pulverized/crushed material into various sizes by using a sieve.

separation: Dividing debris/material into groups of similar materials by manual or mechanical means.

sheepsfoot roller: A cylindrical steel drum to which knob-headed spikes are fastened; used for compacting earth.

shredder: A mechanical device used to break up waste material into smaller pieces by a tearing action.

skid steer loader: A small loader with a front-mounted bucket--may be fueled by gas, diesel, or propane (commonly referred to as a Bobcat).

sludge: A waste material in the form of a high-solids-content liquid--generally produced from water and sewage treatment processes.

slurry: A pumpable mixture of liquid and fine insoluble solids.

sorbent: A material, compound, or system that can provide a sorption function, such as absorption, adsorption, or desorption.

spoil: Excavated rock and soil.

stabilization: A process by which waste is converted to a more chemically stable form (solidification or chemical reaction).

sump: Pit, tank, or reservoir in which water is collected or stored.

surfactant: A surface-active substance (e.g., detergent).

trackhoe: A tracked backhoe

trommel: A revolving cylindrical frame surrounded by wire cloth, open at both ends--used for separation.

V-ditching bucket: A hoe bucket backhoe attachment that reduces cave-in.

water laser: Device that directs and sprays water at very high pressures (>30,000 psi) and can be used as a cutting tool.

winch: Motorized machine with a drum on which to coil a rope, cable, or chain for hauling or pulling.

Copyright Notice

Table 5   From Ware, S. A. and G. S. Jackson. 1978. Liners for Sanitary Landfills and Chemical and Hazardous Waste Disposal Sites. Used by permission of the U.S. Environmental Protection Agency - Center for Environmental Research Information. Cincinnati, OH.

Table 6   From Ware, S. A. and G. S. Jackson. 1978. Liners for Sanitary Landfills and Chemical and Hazardous Waste Environmental Protection Agency - Center for Environmental Research Information. Cincinnati, OH.

Table 7   From Church, H.K. 1981. Excavation Handbook. Used by permission of McGraw-Hill Publishing Co. New York, NY.

Figure 4   From John Deere and Co., Industrial Division. 1990 Used by permission of John Deere and Co., Moline, IL.

Figure 5   From Church, H.K. 1981. Excavation Handbook. Used by permission of McGraw-Hill Publishing Co. New York, NY.

Table 8   From Church, H.K. 1981. Excavation Handbook. Used by permission of McGraw-Hill Publishing Co. New York, NY.

Figure 8   From Perry, R. H. 1984. Perry's Chemical Engineer's Handbook. Reproduced with permission from McGraw-Hill Publishing Co. New York, NY.

Table 9   From U. S. Environmental Protection.Agency 1985. Remedial Action at Waste Disposal Sites, Handbook. Used by permission of the U. S. Environmental Protection Agency - Center for Environmental Research Information. Cincinnati, OH.

Figure 10 From Ware, S. A. and G. S. Jackson. 1978. Liners for Sanitary Landfills and Chemical and Hazardous Waste Disposal Sites. Used by permission of the U.S. Environmental Protection Agency - Center for Environmental Research Information. Cincinnati, OH.

Figure 11 From Crisafulli Pump Company. 1990. Used by permission of the Crisafulli Pump Company, Glendive, MT.

Figure 12 From Bonner, T.A. 1981. Engineering Handbook for Hazardous Waste Incinerators. Used by permission of the U.S. Environmental Protection Agency - Center for Environmental Research Information. Cincinnati, OH.

Figure 13 From Shredding Systems, Inc. 1989. Used by permission from Shredding Systems Inc. Wilsonville, OR.

Figure 15   From Piqua Engineering, Inc. 1989. Used by permission of Piqua Engineering, Inc. Piqua, OH.

Figure 17   From Warren Springs Laboratory. 1989. Used by permission of the Warren Springs Laboratory. Hertfordshire, United Kingdom.

Appendix G

Backhoe attach.
pg. 169
From John Deere and Co., Industrial Division. 1990. Used by permission of John Deere and Co., Moline, IL.

Crane
pg. 170
FromGrove Worldwide Manufacturing Co. 1990. Used by permission of Grove Worldwide Manufacturinf Co., Shady Grove, PA.

Crawler tractor
pg. 171
From John Deere and Co., Industrial Division. 1990. Used by permission of John Deere and Co., Moline, IL.

Dredging system
pg. 172
From Crisafulli Pump Company. 1990. Used by permission of the Crisafulli Pump Company, Glendive, MT.

Barrel handler
pg. 173
From LaBounty Manufacturing, Inc. 1990. Used by permission of LaBounty Manufacturing, Inc. Two Harbors, MN.

Earth auger
pg. 174
From Ditch Witch Sales and Service. 1990. Used by permission of Ditch Witch Sales and Service. Cincinnati, OH.

Excavator (tired)
pg.175
From John Deere and Co., Industrial Division. 1990. Used by permission of John Deere and Co., Moline, IL.

Excavator (tracked)
pg.176
From John Deere and Co., Industrial Division. 1990. Used by permission of John Deere and Co., Moline, IL.

Flexible containers
pg. 177
From Helios Container Systems, Inc. 1990. Used by permission of Helios Container Systems, Inc. Bloomingdale, IL.

Hand-held access.
pg. 178
From Allied Steel and Tractor Products, Inc. 1990. Used by permission of Allied Steel and Tractors Products, Inc. Solon, OH.

Bag liners
pg. 179
From Packaging Research and Design Corp. 1990. Used by permission of Packaging Research and Design Corp. Madison, MS.

Water laser pg. 180	From Allied Steel and Tractor Products, Inc. 1990. Used by permission of Allied Steel and Tractors Products, Inc. Solon, OH.
Hoe-ram pg. 181	From Allied Steel and Tractor Products, Inc. 1990. Used by permission of Allied Steel and Tractors Products, Inc. Solon, OH.
Horizontal auger pg. 182	From Brown Bear Corp. 1990. Used by permission of Brown Bear Corp. Lenox, IA.
Vacuum pg. 183	From Super Products Corp. 1990. Used by permission of Super Products Corp. Milwaukee, WI.
Loader (tired) pg. 184	From John Deere and Co., Industrial Division. 1990. Used by permission of John Deere and Co., Moline, IL.
Loader (tracked) pg. 185	From John Deere and Co., Industrial Division. 1990. Used by permission of John Deere and Co., Moline, IL.
Loader assecc. pg. 186	From John Deere and Co., Industrial Division. 1990. Used by permission of John Deere and Co., Moline, IL.
Mixer pg. 187	From Maxon Industries, Inc. 1990. Used by permission of Maxon Industries, Inc. Milwaukee, WI.
Mobile shear pg. 188	From LaBounty Manufacturing, Inc. 1990. Used by permission of LaBounty Manufacturing, Inc. Two Harbors, MN.
Oil skimmer pg. 189	From Oil Mop, Inc. 1990. Used by permission of Oil Mop, Inc. Belle Chase, LA.
Holding tank pg. 191	From Aero-Tec Laboratories, Inc. 1990. Used by permission of Aero-Tec Laboratories, Inc. Ramsey NJ.
Vibrating screen pg. 192	From Powerscreen of America. 1990. Used by permission of Powerscreen of America. Louisville, KY.
Pumping system pg. 194	From Crisafulli Pump Company. 1990. Used by permission of the Crisafulli Pump Company, Glendive, MT.
Rad'ctive containers pg. 195	From Scientific Ecology Group, Inc. 1990. Used by permission of Scientific Ecology Group, Inc. Oak Ridge, TN.

Copyright Notice

Forklift pg. 196	From Massey Ferguson Industrial Machinery, Inc. 1990. Used by permission of Massey Ferguson Industrial Machinery, Inc. Chamblee, GA.
Shredder pg. 197	From MAC/Saturn Corporation. 1990. Used by permission of MAC/Saturn Corporation. Grand Prairie, TX.
Shredder/ compactor pg. 198	From MAC/Saturn Corporation. 1990. Used by permission of MAC/Saturn Corporation. Grand Prairie, TX.
Shredder/ mixer pg. 199	From Powerscreen of America. 1990. Used by permission of Powerscreen of America. Louisville, KY.
Skid-steer loader pg. 200	From Melroe Company. 1990. Used by permission of Melroe Company. Fargo, ND.
Loader assecc. pg. 201-2	From Melroe Company. 1990. Used by permission of Melroe Company. Fargo, ND.
Trenching machine pg. 203	From Ditch Witch Sales and Service. 1990. Used by permission of Ditch Witch Sales and Service. Cincinnati, OH.

Acronyms/Abbreviations

ACE	--	Army Corps of Engineers
CERCLA	--	Comprehensive Environmental Response Compensation and Liability Act
cc	--	Cubic centimeters
cfm	--	Cubic feet per minute
CFR	--	Code of Federal Regulations
CWA	--	Clean Water Act
cylm	--	Cubic yard loose measurement
cysm	--	Cubic yard struck measurement
DOT	--	Department of Transportation
EPA	--	U.S. Environmental Protection Agency
ERCS	--	Emergency Response Cleanup Services
gpm	--	Gallons per minute
hp	--	Horsepower
HEPA	--	High-efficiency particulate air filter
MSW	--	Municipal solid waste
NCP	--	National Contingency Plan
NPL	--	National Priorities List
NRC	--	Nuclear Regulatory Commission
OSC	--	On-Scene Coordinator
OTR	--	Over-the-road (tractor)
PAH	--	Polynuclear aromatic hydrocarbon
PCB	--	Polychlorinated biphenyl
psi	--	Pounds per square inch
PVC	--	Polyvinyl chloride
RCRA	--	Resource Conservation and Recovery Act
ROD	--	Record of Decision
RPM	--	Remedial Project Manager
SARA	--	Superfund Amendments and Reauthorization Act
SOP	--	Standard Operating Procedure
THC	--	Total hydrocarbons
tph	--	Tons per hour
UST	--	Underground storage tank
VOC	--	Volatile organic compound

TD 1040 .D67 1992

Dosani, Majid.

Materials-handling
 technologies used at
 hazardous waste sites